Water Science and Technology Library

Volume 93

The aim of the *Water Science and Technology Library* is to provide a forum for dissemination of the state-of-the-art of topics of current interest in the area of water science and technology. This is accomplished through publication of reference books and monographs, authored or edited. Occasionally also proceedings volumes are accepted for publication in the series. *Water Science and Technology Library* encompasses a wide range of topics dealing with science as well as socio-economic aspects of water, environment, and ecology. Both the water quantity and quality issues are relevant and are embraced by *Water Science and Technology Library*. The emphasis may be on either the scientific content, or techniques of solution, or both. There is increasing emphasis these days on processes and *Water Science and Technology Library* is committed to promoting this emphasis by publishing books emphasizing scientific discussions of physical, chemical, and/or biological aspects of water resources. Likewise, current or emerging solution techniques receive high priority. Interdisciplinary coverage is encouraged. Case studies contributing to our knowledge of water science and technology are also embraced by the series. Innovative ideas and novel techniques are of particular interest.

Comments or suggestions for future volumes are welcomed.

Vijay P. Singh, Department of Biological and Agricultural Engineering & Zachry Department of Civil and Environment Engineering, Texas A&M University, USA
Email: vsingh@tamu.edu

More information about this series at http://www.springer.com/series/6689

Binaya Kumar Mishra · Shamik Chakraborty ·
Pankaj Kumar · Chitresh Saraswat

Sustainable Solutions for Urban Water Security

Innovative Studies

Binaya Kumar Mishra
School of Engineering
Faculty of Science and Technology
Pokhara University
Pokhara, Nepal

Pankaj Kumar
Natural Resources and Ecosystem Services
Institute for Global Environmental Strategies
Hayama, Japan

Shamik Chakraborty
Faculty of Sustainability Studies
Hosei University
Tokyo, Japan

Chitresh Saraswat
The Fenner School of Environment & Society
Australian National University
Canberra, ACT, Australia

Funder name: Asia-Pacific Network for Global Change Research
Funder ID: https://doi.org/10.13039/100005536

ISSN 0921-092X ISSN 1872-4663 (electronic)
Water Science and Technology Library
ISBN 978-3-030-53112-6 ISBN 978-3-030-53110-2 (eBook)
https://doi.org/10.1007/978-3-030-53110-2

This Springer imprint is published by the registered company Springer Nature Switzerland AG
The registered company address is: Gewerbestrasse 11, 6330 Cham, Switzerland

Preface

How our society responds to the growing problem of urbanization induced environmental changes is a pressing issue with urban sustainable development. One way to understand this issue is through the case of water, one of the most precious resources in urban areas. Water resources face widespread deterioration due to unsustainable urbanization. The fundamental growth and development of human societies are dependent on water resources, and so the ecosystems on which human population directly depend on. In fact, an increasing number of researchers and professionals are now engaged in finding out solutions to this vast problem with huge social implications. However, the results largely remain not practiced on the ground and urban areas continue to adversely affect freshwater environments. But the interest in studying urban areas more with successful case studies, capturing the diversity of possibilities associated with urban water management—from increasing and protecting biodiversity in urban ecosystems to engineering solutions that are based on long-term benefits—is on the rise. This book captures notable sustainable solutions backed up by robust theoretical and scientific background analyzed as keys to these successful practices.

The foundations of the book were laid through the project Asia Pacific Network for Global Change Research (APN), titled "Climate Change Adaptation through Optimal Stormwater Capture Measures: Towards a New Paradigm for Urban Water Security" (Project ID: ARCP2014-20NMY-Mishra). Another major factor has been with the author's interests in understanding and explaining urban water resource problems to professionals (academicians, policy-makers) as well as common citizens, thereby address and better equip water resource management policies in urban areas. The chapters in this book cover a wide range of issues regarding urban water security, with a focus on various success stories. We concentrate on success stories in detail mainly as we think they give hope for the future generations and provide examples, and showcase our ability to adapt in difficult situations such as degradation of a most vital resource. The most important feature of this book is the theoretical and conceptual viewpoints backed up by case studies of real-world situations, which mainly came through our experiences in the field of water resource management. Throughout the book, it is emphasized that sustainable water resource management in urban areas plays a pivotal role in a broad range of issues relater to water security such as social stability, economic and environmental well-being. The book thus can also be taken as a handbook of workable answers to the problem of not only urban water resource management alone but also a nexus that relates urban sustainability through water issues.

Finally, we hope that this book will be an important step toward informing urban policy decisions regarding water security issues. We also hope that the combination of these valuable steps, together with other similar studies, will open new pathways for the urban society to deal with sustainable water resource management.

Pokhara, Nepal Binaya Kumar Mishra
Tokyo, Japan Shamik Chakraborty
Hayama, Japan Pankaj Kumar
Canberra, Australia Chitresh Saraswat
July 2020

The original version of the book was revised: The funder information has been added. The correction to the book is available at https://doi.org/10.1007/978-3-030-53110-2_10

Acknowledgements

We owe an enormous debt of gratitude to the reviewers who gave us detailed, constructive comments and suggestions on one or more chapters, including Dr. Srikantha Herath, Dr. Abhik Chakraborty, Dr. Chandan Banerjee and Dr. Malcolm J. M. Cooper. Also, we are particularly grateful to our families and friends, Mr. Shyam Narayan Mishra, Mr. Dhruvesh Saraswat, Dr. Anil Kumar Gupta, Mr. Kiran Kovuru, Mr. Vivek Sharma, Mr. Aadesh Saraswat, Ms. Yukti Sharma, Ms. Divyanshi Sharma, Mr. Bhavesh Patel, Mr. Shishir, Mr. Vikas Agarwal, Mr. Deepak Mohta, Mr. Ajeet Pal, Dr. Prateek Nigam, Mr. Sabyasachi Raj Kumar, Mr. Sunil Raiyani and Mr. Sonil Singh, who took time out of their busy schedule for reading the different versions of draft, devoting time to converse on simplifying jargons and concepts in the book, exploring particular facets and dicussing the rationales of the book. The funding of this book is supported by the project under Asia Pacific Network for Global Change Research (APN), titled 'Climate Change Adaptation through Optimal Stormwater Capture Measures: Towards a New Paradigm for Urban Water Security' (Project ID: ARCP2014-20NMY-Mishra).

This book has been successful due to the Water and Urban Initiative (WUI) research program at United Nations University Institute for the Advanced Study of Sustainability (UNU-IAS), where we conceptualized the idea after debating hours on the current research gap and situations we learned during working at different projects. There is great amount of exposure of discussion and work with the colleagues, work and administration at UNU-IAS who helped us to refine our thinking. Most importantly, the programs, teaching and discussion at the University of Tokyo Integrated Research System for Sustainability Science (IR3S) 1, Japan, the Japan Society for the Promotion of Science (JSPS), School of Engineering, Pokhara University, Nepal, The Fenner School of Environment & Society, Australian National University, Canberra, Australia, Department of Sustainability Studies, Hosei University, Japan, and Natural Resource and Ecosystem Services, Institute for Global Environmental Strategies (IGES), Japan. Without their support, this book would not have been a reality, especially as a lot of the information and case studies discussed in the book come from the authors' research works in these institutions. The authors would like to thank all the people involved at Springer for making this project a success, particularly Dr. Petra van Steenbergen, Ms. Amudha Vijayarangan, Ms. Margaret Deignan and Ms. Bhagyalakkshme Sreenivasan, who consistently supported us throughout the writing process.

Finally, we want to thank our wives Ms. Ranjita Jha, Ms. Jyoti Porwal, Ms. Yumi Chakraborty and Ms. Mai Takahashi Saraswat, for their inspiration, continuous support and encouragement to finish this project. Last but not least, we would like to thank our parents and families who supported us in thick and thin and kept us motivated to complete the book.

Contents

1 Urban Water Security: Background and Concepts 1
1.1 Background 1
1.2 Water Security: Concept and Evolutions 5
1.3 Water Security in Global Change Context 8
1.4 Water Security Assessment and Indicators 9
1.4.1 Associated Factors and Approaches 12
1.4.2 Sustainable Water Management 15
1.5 Summary 18
References 20

2 Urban Water Security Challenges 25
2.1 Background 25
2.2 Urbanization 27
2.3 Climate Change 29
2.4 Implications on Water Security 30
2.4.1 Hydrological Cycle 30
2.4.2 Water Shortage 31
2.4.3 Land Subsidence 31
2.4.4 Surface and Groundwater Pollution 32
2.4.5 Human Health 32
2.4.6 Ecosystem and Biodiversity 33
2.5 Case Studies 33
2.5.1 Bagmati River Flood, Nepal 33
2.5.2 Ciliwung River Flood, Indonesia 35
2.6 Summary 38
References 39

3 Urban Water Demand Management 41
3.1 Background 41
3.2 Current Perspective and Prospects 42
3.3 Urban Water Demand Management Strategies 45
3.4 Case Study: Effectiveness of Urban Water Demand Management Strategies in Kathmandu Valley, Nepal 47
3.4.1 Introduction 47
3.4.2 Methodology 50
3.4.3 Supply- and Demand-Side Management Strategies Formulation and Analysis 50
3.5 Discussion and Summary 52
References 54

4 Water Quality Restoration and Reclamation 59
4.1 Background 59
4.2 Effects of Land-Use Change 61
4.3 Water Quality and Health Nexus 63
4.4 Wastewater: Global Trends 64
4.5 Wastewater, Sanitation and the Sustainable Development Agenda 67
4.6 Innovations in Water Quality Treatment 68
4.7 Wastewater Reclamation Technologies 69
4.8 Case Study: Sustainable Water Environment Management in Jakarta 74
4.9 Summary 77
References 79

5 Landscape-Based Approach for Sustainable Water Resources in Urban Areas 83
5.1 Background 83
5.1.1 Effects of Urbanization on Water Resources: Some Notable Examples 84
5.2 Water and Landscape Issues 86
5.3 The Role of the Ecosystem Service Approach to Water Security 89
5.3.1 Ecosystem Service Components that Can Better Address Urban Water Security 91
5.4 Landscape Diversity as a Tool for Urban Water Management 93
5.4.1 Areas Inside the Urban Domain 93
5.4.2 Areas Outside the Urban Domain 94

5.5 Natural Disturbance Regimes in Urban Water Security 95
5.6 Case Studies 96
5.6.1 Small-Scale and Holistic Landscape Conservation in Town Planning: The Case of Aya, Japan 96
5.6.2 Prioritizing Diverse Stakeholder Knowledge for Water Resource Governance in Urban Areas 100
5.6.3 Multiple Provisioning Services from Peri-Urban Watershed Environments: Case Study of Jala-Jala Watershed Near Manila 102
5.7 Summary 106
References 108

6 Urban Stormwater Management: Practices and Governance 115
6.1 Background 115
6.2 Global Overview 116
6.2.1 Urbanization 119
6.2.2 Climate Change 121
6.3 Regulatory Policies 121
6.4 Tools and Approaches in Optimal Stormwater Management 123
6.4.1 Historical Trends 124
6.4.2 Potential Role of Remote Sensing and GIS Technologies in Stormwater Management 125
6.4.3 Numerical Simulation Modeling 127
6.4.4 Economic Assessment 127
6.5 Urban Stormwater Governance: Case Studies 130
6.5.1 Tokyo 130
6.5.2 Bangkok 132
6.5.3 Hanoi 134
6.6 Summary 137
References 141

7 Numerical Modeling and Simulation for Water Management 147
7.1 Background 147
7.2 Model Classifications 148
7.2.1 Black Box Models 149
7.2.2 Conceptual Models 149
7.2.3 Physically Based Distributed Models 150
7.3 Model Selection 150
7.4 Calibration–Validation 151
7.5 Quantifying Uncertain Future 153
7.6 Open Access Data and Software 154

7.7 Applications of Remote Sensing and GIS 155
7.8 Case Studies 156
7.8.1 Simulation of Flood Inundation in Manila Using FLO-2D Model 156
7.8.2 Use of RS and GIS for Assessment of Land-Use Change Impacts on Runoff 158
7.9 Summary 158
References 160

8 Urban Water Governance: Concept and Pathway 161
8.1 Background 161
8.2 Water Governance: South Asia and India 162
8.3 Urban Water Governance: Conceptualization 163
8.4 Urban Water Governance Pathway: Adaptive, Polycentric and Hybrid 165
8.5 Barriers to Implementation: Indian Context 167
8.6 Summary 169
References 171

9 Toward Sustainable Solutions for Water Security 175
9.1 Background 175
9.2 Long-Term Sustainable Solutions 176
9.3 Interplay Between System Parts 177
9.4 Conservation for Water Security: A Compound Criterion 178
9.5 Making Meanings of Large Data and Stress on Pragmatic Solutions 178
9.6 Ways Forward and Gaps Remaining 180
9.6.1 Political Bottleneck and Lack of Integration 180
9.6.2 Education and Lifestyles 180
9.6.3 Channelizing Research and Innovation into Policies More Efficiently 181
9.6.4 Data Creation and Management 181
9.6.5 Understanding on an Individual Basis/Leadership 181
References 182

Correction to: Sustainable Solutions for Urban Water Security C1

About the Authors

Binaya Kumar Mishra is currently engaged as a full-time professor at the School of Engineering, Pokhara University, Nepal. Prior to this, he worked as a research fellow at the United Nations University, Institute for the Advanced Study of Sustainability (UNU-IAS), Tokyo. He is involved in research and teaching activities for Bachelor (Civil Engineering) and Master (Hydropower Engineering; Public Health and Disaster Engineering) programs. His research and teaching interests include water resources management; climate and ecosystem change adaptation; hydrologic and environmental modeling and applications of GIS and remote sensing. Prior to joining Pokhara University, he briefly worked as an associate professor at the Central Campus of Engineering, Mid-Western University, Nepal. Earlier, during October 2009–March 2018, he worked as a researcher and faculty member at different academic institutions in Japan. He also worked as an irrigation engineer at the Ministry of Irrigation; a senior lecturer at Kathmandu Engineering College, Tribhuvan University, and a consulting engineer at Kathmandu, Nepal, during 1999–2006. His research works have been published in several books, journals, reports and proceedings.

Shamik Chakraborty is a lecturer at the Department of Sustainability Studies, Hosei University, Japan. Prior to this, he has worked as a JSPS-UNU postdoctoral research fellow at United Nations University, Institute for the Advanced Study of Sustainability (UNU-IAS), Tokyo, and a visiting research fellow at the Integrated Research System for Sustainability Science (presently, Institute for Future Initiatives) at the University of Tokyo. As a human geographer, he is interested in studying human-environment interactions in coastal and river basin environments from a social–ecological systems point of view. He has worked with the concepts of social–ecological systems, indigenous and local knowledge in diverse ecosystems in Japan, India, Nepal, Bangladesh and the Philippines. He is currently involved with applying ecosystem services concept as an anthropocentric tool to understand the contact points of human society with diverse sets of benefits from the inland, river basin and coastal environments for biodiversity conservation, climate change mitigation and sustainable development. He has published his works in a number of international academic journals and edited volumes.

Pankaj Kumar is currently working in the Institute for Global Environmental Strategies (IGES), Japan, as a senior policy researcher in the field of water resources and climate change adaptation. Prior to this, he worked as a research fellow at the United Nations University, Institute for Advanced Study of Sustainability (UNU-IAS), Tokyo, for four years. Recently, his research work focused on 'hydrological simulation and scenario modeling for clean urban water environment in South-East Asian developing cities' a transdisciplinary work aimed to enhance community resilience to global change and provide policy-relevant solutions. He holds a doctoral degree in Geo-Environmental Science from the University of Tsukuba, Japan. In parallel, he had work experience as Chapter Scientist for Working Group-II of Fifth Assessment Report (AR5) of IPCC. Currently, he is also working as a scoping expert for nexus assessment for IPBES. He has several peer-reviewed journal articles and chapters of international repute to his credit.

Chitresh Saraswat is an interdisciplinary researcher and doctoral scholar in the Fenner School of Environment & Society, Australian National University, Australia. Trained as an environmental geographer and computer scientist, he focuses on accelerating transitions toward water sustainability in the Global South. His research interests span across water governance, innovation and sustainability transitions. He has over ten years of research and industry experience, both nationally and internationally. He has developed this through roles with United Nations agencies, energy and water utility and corporate (IT) sector firms. He received a Master of Science degree in sustainability from the United Nations University, Institute for Advanced Study of Sustainability (UNU-IAS), Tokyo, and Joint Diploma (sustainability science) from Graduate School of Frontier Sciences, the University of Tokyo and UNU-IAS, Tokyo, Japan. He has published several peer-reviewed outputs in journals with international repute and books in the research areas of sustainability, water management, security and governance and cyber-physical systems.

Chapter 1
Urban Water Security: Background and Concepts

1.1 Background

Currently, more than 1.1 billion people have inadequate access to clean drinking water globally, and approximately 2.6 billion people lack basic sanitation (Pink 2016; Jain 2012). Water is the foundation of life and a basic necessity for everyone, but lack of access is gradually becoming a crisis for millions of people around the world that is responsible for poor health, destruction of livelihoods and unnecessary suffering for the poor (Hanjra and Qureshi 2010). Therefore, overcoming the water crisis is one of the greatest challenges faced by our generation (IPCC 2007), and developing clean potable water, efficiently managing wastewater and providing basic sanitation facilities are the foundations for sustainability and human progress (UN-Water 2010; Tremblay 2010). Successfully achieving these goals would catalyze progress in many sectors such as public health, energy security, climate resilience and poverty reduction, as well as accelerating the pace toward achieving the Sustainable Development Goals (SDGs), which were recently approved in the 71st Session of the United Nations General Assembly (Sachs 2012). From the Millennium Development Goals (MDGs) to the SDGs, the focus of the water security concept has shifted from only water supply and demand in cities toward the perception of water as an economic resource shared between countries (Connor 2015). This shift also emphasizes the concept of water governance, inlcuding its capacity to manage water efficiently and equitably (Conca 2006; Gareau and Crow 2006). Therefore, the definition of water security is rapidly changing to include ensuring every person has reliable access to sufficient safe water at an affordable price to enable a healthy and productive life, as well as maintaining water-related ecological systems for future generations (Cook and Bakker 2012). However, water stress is rapidly increasing in the developing parts of the world, including countries in Sub-Saharan Central Africa, Western South America, Australia, Southeast Asia, China and the world's most water-stressed region, the Middle East, where the water availability per person is less than 1100 cubic meters. In the past several years, the water situation has improved, but at least

B. K. Mishra et al., *Sustainable Solutions for Urban Water Security*, Water Science and Technology Library 93,
https://doi.org/10.1007/978-3-030-53110-2_1

650 million people still lack access to clean drinking water. The rapidly developing regions of Asia, Africa and South America face groundwater overexploitation and skewed water supply due to explosive population growth and negative impacts from climate change (Cook and Bakker 2012). Overall, the Asian region is also one of the most rapidly urbanizing parts of the world, with the urban population growing at the alarming rate of 2.3% annually, which is greater than the global average of 2% (Cohen 2006). Approximately, 16 megacities with populations of 10 million inhabitants or more, more than half of the world's megacities, are located in this region (UN DESA 2014), and it is estimated that the urban population of the continent of Asia will exceed the rural population by the year 2022 (ADB 2008). Recently, the Asian Water Development Outlook Report by the Asian Development Bank predicted that more than 1.7 billion people worldwide will lack access to basic sanitation by the year 2050; approximately 3.4 billion people will be living in water-stressed areas (AWDO 2016); and water demand will increase by 55% (Chellaney 2011).

Figure 1.1 shows the future global water stress scenario by the year 2040; most of Australia, Northern and Southern Africa, the USA and Western South America will face water stress in the form of a high withdrawal to supply ratio (40–80%) (WRI 2015), and a large part of Asia will face extremely high (>80%) water stress. Managing water scarcity to this extent requires new definitions of water security that emphasizes the importance of maintaining the ecological systems that provide water (Wagener et al. 2010) and increasing the significance of water governance in solving the problem. In this scenario, it is very important to design a wide range of sustainable water security solutions that address the multiple instances of increasing water scarcity around the globe and effectively address local constraints. Figure 1.2 indicates the urgency and necessity for a range of sustainable solutions by showing

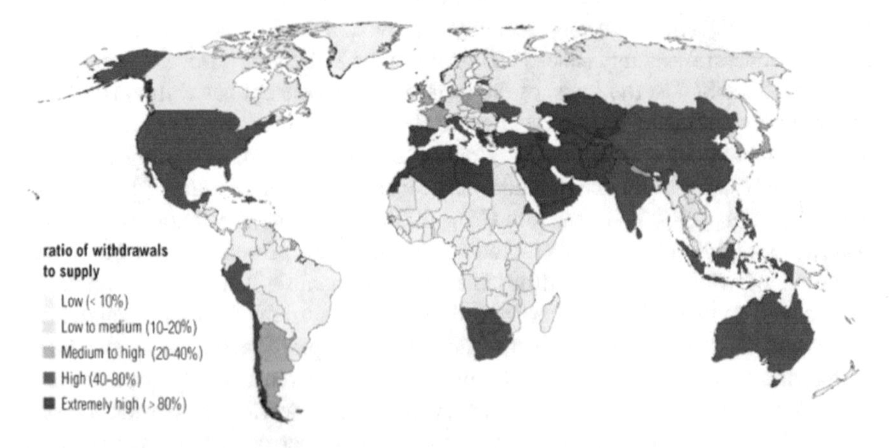

Fig. 1.1 Projected country-level water scarcity for the year 2040. Projections are based on a business-as-usual scenario using SSP2 and RCP 8.5. *Source* WRI 2015

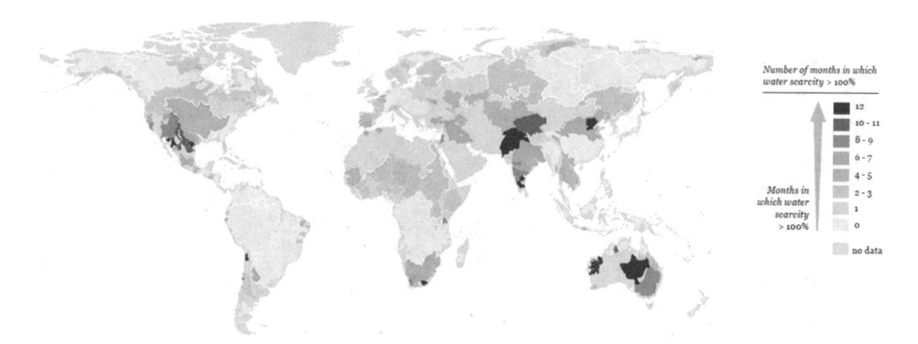

Fig. 1.2 Number of months in which severe water scarcity was observed in more than 400 river basins from 1996 to 2005. *Source* WWF 2012

the number of months in which water scarcity reached 100% from 1996 to 2005 in more than 400 river basins around the world (Fig. 1.2) (WWF 2012). Thus, there is an urgent need for a sustainable solution to achieve water security based on local considerations. The Southern USA, the Western coast of South America, most of South–Eastern Australia, North–Western India, Pakistan and other nearby regions are already facing the highest degree of water scarcity; therefore, localized sustainable solutions to reduce water stress are needed immediately.

One important dimension of effective sustainable water solutions is dealing with societies that have persistent water inequalities, the growing development divide and the competition for scarce resources. The inhabitants of many developed nations, as well as high-income regions and cities, enjoy the delivery of several hundred liters of water per day at a very low price, but in other parts of the world, poor households in both rural and urban areas do not have an adequate supply of safe drinking water (UNICEF and WHO 2011). Similarly, within the agricultural and industrial sectors, small famers have the least access to water in wealthy areas. Jenson (2016) recommended solving these water issues through the development of equitable water policies, investment in water infrastructure, public–private partnerships and institutional approaches (Jenson 2016), and the concept of strong and adaptive water governance was effectively explored by Huitema et al. (2009), who defined adaptive water management comprising four institutional prescriptions. To manage water resources, there is first a strong need for a collaborative governance system along with a high level of public participation in decision-making. The authors recommended an experimental approach toward water-related resource management and highly emphasized water management at the regional scale, but they also raised concerns about the complexities associated with collaboration and participation in the management of water resources, the problem of real-world experimentation and the inadequacy of governance structure at the regional level (Huitema et al. 2009; Lankford et al. 2013).

UN report on global water quality perspective based on changes in the nitrate concentrations at various river mouths during 1990–1999 and 2000–2007 is shown

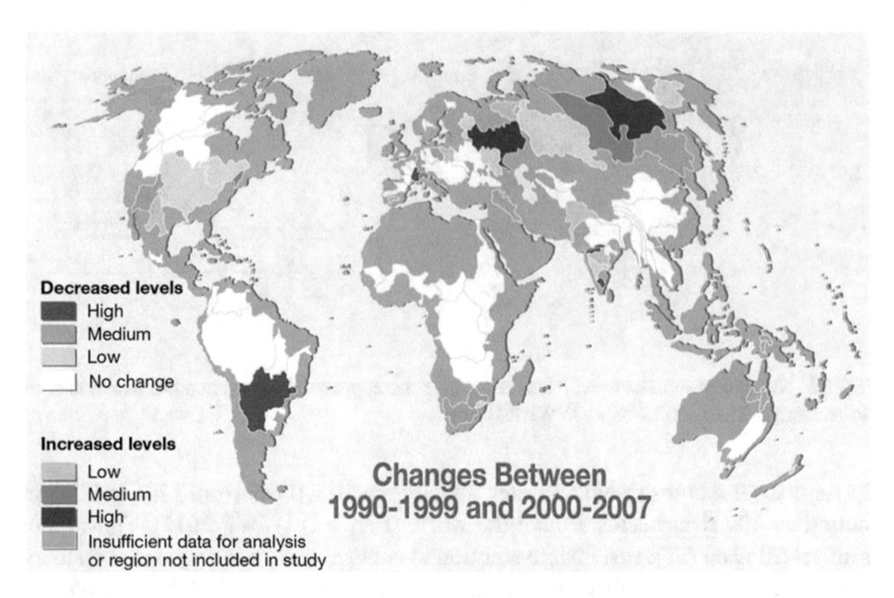

Fig. 1.3 Water quality issues around the world: nitrate concentrations at the mouths of rivers. Adapted from the United Nations Environment Program (UNEP); The Water Program 2008, and the National Water Research Institute, Environment Canada, Ontario 2008. *Source* United Nations Environment Program (UNEP) (2016)

in Fig. 1.3. Here, the results raise concerns over another dimension of water security, which is related to water quality or water pollution and the complexities of designing sustainable solutions to deal with stormwater, wastewater, water quality and the pollution of freshwater (Gleick and Ajami 2014).

Most of the wastewater and stormwater generated in cities are discharged with 85–90% of its full load of pollutants and toxic compounds (Henze et al. 2001), and high concentration of contaminants/pollutants here is primarily responsible for degrading the quality of both surface and groundwater resources as well as the associated coastal regions (Winter et al. 1998). Furthermore, water pollution affects the economic future of communities and puts human well-being at risk. Therefore, a new water security paradigm must consider treated wastewater as a resource rather than a form of waste (Asano 2002).

It is widely acknowledged that holistic approaches to address water security challenges are needed that include social, economic and environmental dimensions at multiple scales (WSSD 2002). This chapter explores a range of water security dimensions while capturing the dynamic and constantly evolving paradigms related to the subject, and it offers a holistic outlook for addressing water challenges by recommending a portfolio of sustainable solutions to geographically distributed water security challenges. The premises of this study are based on the analysis of case studies from different geographical regions, and it is argued that water security is directly related to the growth of a country's gross domestic product (GDP) or in other words, its economy. Thus, water management and economic growth are strongly related.

Therefore, investing in effective water management can be meaningful for long-term returns regarding economic growth and poverty reduction.

1.2 Water Security: Concept and Evolutions

Water scarcity and water security are interdependent; the problem of water scarcity and stress creates water insecurity, so resolving water scarcity issues represents a step toward achieving water security in any region. Water scarcity can be natural but also anthropogenic; there is enough freshwater on the planet for the entire population, but uneven distribution and its contamination make our water supply insecure. To achieve water security, it is necessary to understand the types of water scarcity and how to overcome them. Water scarcity can be classified into two major types: physical and economic. Physical water scarcity involves water resources that are inadequate to meet demand, including the needs of ecosystems to function effectively; the arid regions of the world must address physical water scarcity. However, water scarcity can also occur due to abundant water resources but not managed effectively, which results into environmental degradation, groundwater depletion and water overexploitation. Economic water scarcity is caused by inadequate water management or planning capacity to satisfy the water demand in a region, and it is estimated to affect more than a quarter of the global population due to inadequate water infrastructure (UNDESA 2016). Inadequate water infrastructure is widespread in underdeveloped or developing regions, where critical conditions for regular water supply often arise for economically poor communities. Increased water consumption is related with increasing income (Sergiusz 2015; Gray and Shadoff 2007); therefore, water-secure and insecure countries are dependent on the cumulative investments in water infrastructure and the related institutions (Fig. 1.4). This dependence implies that more investment in water-related infrastructure will be needed for a water-secure future, and as per the above argument, this investment in water security will bring positive changes to the GDP per capita of a country or region, leading it on the path toward a sustainable future. As illustrated in Fig. 1.4, the water security S-curve described by Grey and Sadoff (2007) shows that water-related risks are high and increasing in developing countries, which are more often water insecure, and this affects the resources available for investment and governance. When resources are made available for management, the benefits are high, and risk is rapidly reduced. As risks lessen and the curve levels off, large new investments are no longer justified and the country is water secure. It is important to note that even in a water-secure country, maintenance plays a vital role; without it, tendency to falling back into the water-insecure state is easier.

In the present context, water security paradigms have become more diverse, expanding from water demand and supply to water quality, ecological concerns and overall human well-being (Cook and Bakker 2012). However, many disciplines tend to focus on different scales within a very narrow perspective. For example, development studies largely deal with national scales; economists are mainly concerned

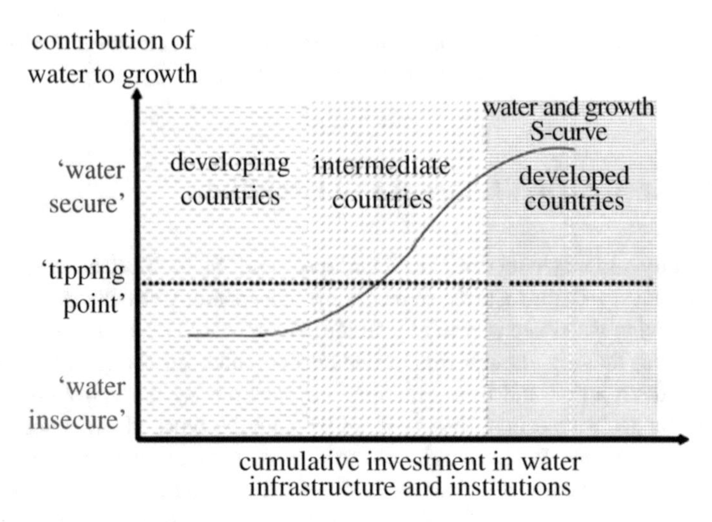

Fig. 1.4 Water security S-curve: the relationship between water and sustainable growth. *Source* Grey and Sadoff 2007

with the economic aspects of a problem; hydrologists often only focus at watershed scales; and social scientists primarily perform research around the community (Cook and Bakker 2012). The new dimensions needed to attain sustainable water security require a longer-term vision that involves a range of sustainable solutions based on economic, environmental and societal aspects according to region-specific needs (Jiang 2015), and the historical evolution of water security technologies is important for understanding the present and future versions of the concept of water security. This concept can be divided into four categories, including Water 1.0, which is the first generation of water technology that includes the aqueducts of the Roman Empire, whose complex web of pipes and channels were used to supply freshwater and transport waste or sewage out of cities. This property remained the norm until the nineteenth century, and the side effect of direct sewage was many water-related diseases, such as cholera and typhoid. In the twentieth century, Water 2.0 dealt with these issues as cities became more crowded and the focus shifted to treating potable water, usually with filtration and/or chlorine, but the sewage continued to pour into rivers, ultimately entering nearby bays or the surrounding sea. The problems of untreated water such as the smell and filth entering downstream cities led to Water 3.0, by which the enormous infrastructure of sewage treatment plants was developed to manage water in an efficient way. In the Water 4.0 era, water technologies deal with water supply and demand supply along with wastewater treatment, and fourth-generation water systems feature water recycling plants that capture and reuse rainwater that would otherwise become runoff and pollute the water bodies in and around cities (Garrick and Hall 2014).

As shown in Fig. 1.5, urban regions require new strategies to cope up with the transitions between water requirement phases as they develop. Water supply cities, which require a water supply to manage demand, evolve in the direction of water-sensitive

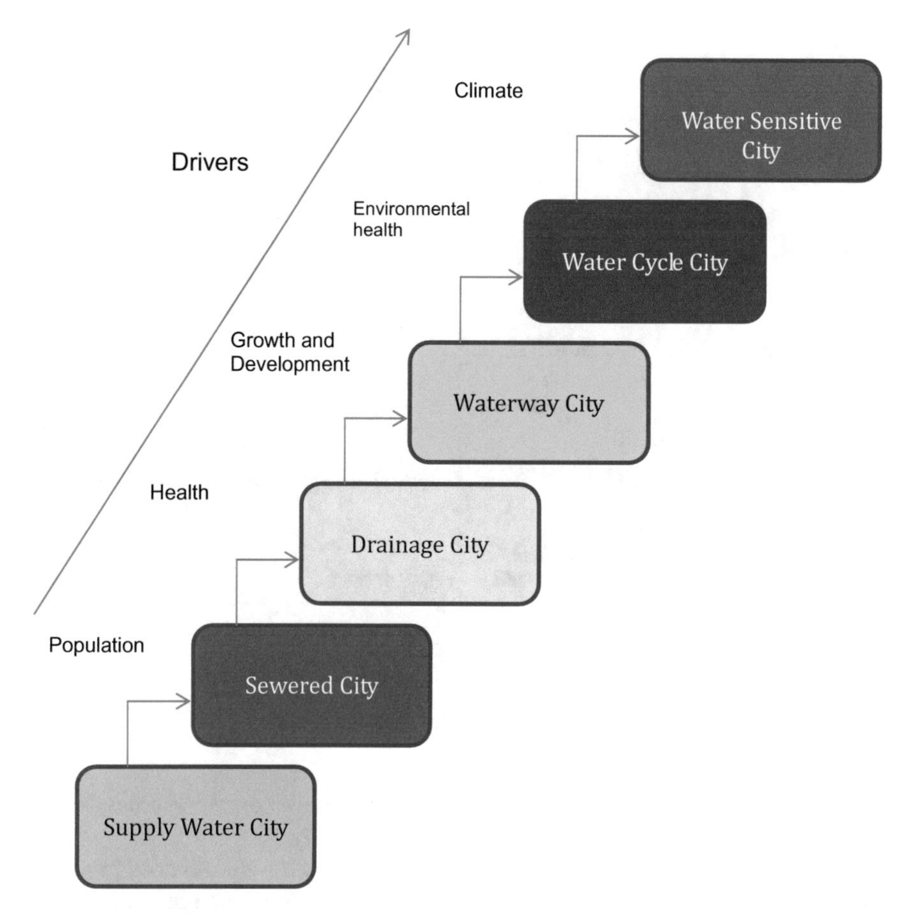

Fig. 1.5 Changing water security paradigms and transition phases. Adapted from Brown et al. (2009)

cities due to various factors, and different drivers are responsible for these water transition phases, such as rapid population growth. To solve the emerging problem of water security, it is important to adapt management responses to the changing context, and as explained by Brown et al. (2008), management strategies emerge in response to social and political drivers (Brown et al. 2009). In addition to historical development, the continuum defines future states of emerging socioeconomic drivers to describe upcoming water security challenges. The water security model may be a simple extrapolation of security and insecurity to a complex nth-dimensional abstract space. Furthermore, water security and insecurity can be illustrated by a tolerable level of water risk at any scale and for any factor, assuming sufficiency and equity for everyone (Cashman 2014; Grey and Garrick 2012). The two-dimensional water security model of Lankford termed the four conditions of water security 'incodys', where

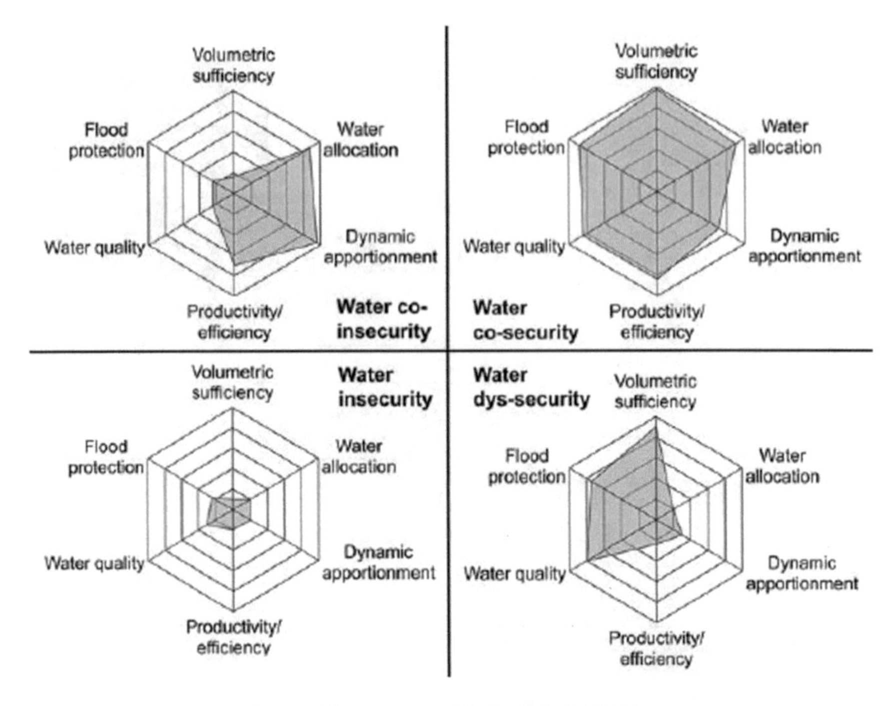

Fig. 1.6 Four water security conditions proposed by Lankford (2013)

'in' represents insecurity; 'co' represents both collective security and coinsecurity; and 'dys' represents 'dys-security' or inequitable security.

Figure 1.6 utilizes a two-dimensional field to illustrate water security based on sufficiency and equity. For example, it can be deduced from the graph that water insecurity is a condition of poorly shared and insufficient water resources that arises because there is not enough supply to meet demand; the water quality is poor, which is associated with health risks and impacts; and/or there are high risks of weather-related water disasters.

1.3 Water Security in Global Change Context

The definition of water security and how communities perceive their water problems are evolving as the complexities of the concept of water security are unraveled. Globally, freshwater resources are under increasing pressure due to urbanization, increasing energy demand, changing lifestyles, increasing population growth and climate change. Several definitions of water security can be found among academics, in policy and in public debate. Water security is more than simply providing sufficient water for people and economic activities; it involves promoting human well-being and healthy aquatic ecosystems as well as increasing resilience to different kinds of water-related disasters. Currently, water security requires that leaders, decision-makers, practitioners and societies continue their efforts/journey toward an integrated approach to water resources management (GWP 2015), and regardless of location, platform for sustainable water management will be built on knowledge of success stories and lessons learned from past experiences. In a constantly changing context, increasing water security calls for a sustainable and adaptive management process that provides a more tailored and systematic response to the strengths of a region, which requires a more holistic and quantitative assessment of water resource management and applies societal, environmental and economic approaches in the proposed solutions.

In the past decade, water security has been increasingly used to more explicitly state the goals of better water management, which has resulted in a range of approaches, definitions and measurements. These goals are based on the natural resource endowment, stage of socioeconomic development and vision for the future of each country, but increasing pressures on water resources have created tensions among different users and are of serious global concern. According to the United Nations (UN-Water 2007), global water use has been growing at twice the rate of population growth over the last century, and nearly 1.2 billion people, or approximately a fifth of the world's population, live in areas subject to physical water scarcity, where there is simply not sufficient water to meet demand (Majumder 2015). Another 1.6 billion people are facing economic water scarcity; they do not have the financial means to access existing water sources. Achieving water security is a necessary condition for achieving the UN Sustainable Development Goals for 2030, which will not be possible without first realizing a water-secure world. The 'UN Agency on Water' defines water security as the capacity of a population to safeguard sustainable access to adequate quantities of water of acceptable quality for sustaining livelihoods, human well-being and socioeconomic development, for ensuring protection against water-borne pollution and water-related disasters and for preserving ecosystems in a climate of peace and political stability (UNU 2013). The United Nations defines water security as safeguarding the sustainable access to adequate quantities of water of acceptable quality for sustaining livelihoods, human well-being and socioeconomic development, ensuring protection against pollution-related water disasters and for preserving ecosystems in a climate of peace and political stability (UN-Water 2013). Water security encapsulates multifaceted and interconnected challenges with a focus on achieving a sense of security, sustainability and human well-being, and the contributing factors are infrastructural, institutional, political, social and, most importantly, financial. Water security is between communities and borders (Tables 1.1 and 1.2).

1.4 Water Security Assessment and Indicators

Spatial and sectoral considerations greatly impact measures of water stress, which is a major determinant of water security. Rijsberman (2006), explained that the supply

Table 1.1 Water supply service performance indicators for urban regions in Asia. Adapted from JICA (2014)

Country	Indonesia	Thailand	Vietnam	Malaysia	Nepal	Singapore
City	Jakarta	Bangkok	HMC	Kuala Lumpur	Kathmandu	Singapore
Population (millions)	8.7	7.958	5.97	1.493	2.8	4.73
Coverage (%)	62	93	84	100	21	100
Water use rate (LPCD)	95	140	121	314	135	220
NRW (%)	50	29	41	35	35–40	4
Year	2008	2009	2009	2008	2015	2009

Table 1.2 Policy recommendations for improved water security. Adapted from OECD (2013)

Category	What to do	Why
Economic and legal	Introduce progressive pricing policies Cross-sectoral management and capacity development Improve institutional and legal frameworks	Create eco-efficient community in terms of water use Reduce vulnerability to water stress Achieve social, economic and environmental goals
Social	Increase data and knowledge Mobilize human and financial resources (social capital) Empower local water users to take greater responsibility Strengthen resistance to water-related risks and hazards	Make informed WRM decisions Create resilient society regarding WRM Improve water governance Increase societal resilience to water-related natural disturbance regimes
Natural	Conserve vital freshwater ecosystems Consider river basins and aquifer systems as biogeographic units for WRM	Maintain the resilience of natural systems Achieve ecosystem-based management

problem occurs when water is truly scarce in the physical sense, and the demand problem occurs when there is sufficient water available that is not used in a sustainable way (Rijsberman 2006). Weather-related water disasters, such as floods and droughts, can occur at the same location within the same year; therefore, determining the average amount of available water to provide sufficient information to measure water scarcity is a major challenge. In the monsoon season, many regions in Asia suffer from water scarcity despite the average annual water resource availability appearing to be abundant. An approach to solving this problem is to build dams to capture runoff and flood water, reduce flood damage and store water for use during the dry period. In large countries such as India, simultaneous water scarcity in the north river basin along with massive flooding in the Southern states shows the complexities of water security problems. At the global scale, current withdrawal of water resources is well below the renewable limit, but the concern is the high unpredictability of water resources in the coming years (Oki and Kanae 2006). Generally, river discharge is unpredictable in smaller river basins, and daily discharge is less predictable than monthly discharge. Furthermore, runoff during floods or wet seasons cannot be used during other periods unless adequate water storage systems are in place. Virtual water trading can be an effective option to address water scarcity, which is estimated to be around 1000 km3/year internationally, although only a very small part of the virtual water market is utilized to compensate for water shortages. Water quality is another major variable in water resource assessment, and there is a need to analyze the level of water quality that is needed against the quality of the available water. This analysis is beneficial for environmental security, human use and understanding water stress, so understanding water scarcity is vital as it affects the views of the most effective policies to address the water crisis.

At global, national and regional levels, countries have developed data, knowledge and methods to reduce water scarcity impacts, but unfortunately, these resources are not being implemented at large scales. In particular, the data collected during research and development projects often do not pragmatically survive after project completion, so there is a need to share results with and build capacity to mitigate water scarcity issues among policy-makers, planners and stakeholders. Hering (2015) emphasized that the full and open exchange of information is vitally important to the sustainable management of water resources; managing the trade-offs among different water uses requires a steady flow of reliable and easy-to-access or open-source data. At present, the data on parameters such as precipitation, river flows, groundwater levels and water quality are collected by many agencies and organizations using different tools, such as remote sensing, and resources such as digital elevation models as well as land use/land cover, precipitation, temperature and soil moisture data are important assets for researchers and decision-makers analyzing water issues. Furthermore, remote sensing or satellite-based data represent leapfrog technologies for developing countries that lack established on-the-ground monitoring networks. This huge amount of information on water resources vastly increases the potential for both research and evidence-based resource management.

Open access tools are available to enable a high level of knowledge sharing including Hydrological Predictions for the Environment (HYPE), a widely tested open-source community of scientists, authorities and consultancies conducting hydrology research, hydrological modeling and source code development. The HYPE tool provides public access to a hydrological model with high operational potential and encourages hydrologists with programming expertise worldwide to contribute to its improvement. Open-source communities ease data sharing and enable ensemble modeling by providing free access to source codes. The study by Malve et al. (2012) employed large amounts of open access data at the European scale to estimate the agricultural non-point pollution load (Malve et al. 2012). The European Environment Agency (EEA) assembles water data from a representative subsample of national monitoring results that countries voluntarily report to the agency every year. In the study, the areas and lengths of the river channels in the watersheds were calculated from the river and catchment database of a joint research center, and the slopes, units, temperatures and pollution loads from the scattered settlements and from the point sources were calculated from WaterGAP3 data layers (Malve et al. 2012).

Other popular open access data platforms that play important roles in managing water resources include the Global Earth Observation System of Systems (GEOSS), the Spatial Information Platform (SIP) under the SWITCH-ON project, the Global Runoff Data Center (GRDC), the Data Distribution Center (DDC) of the Intergovernmental Panel on Climate Change (IPCC) and the Center for Earth Resources, Observation and Science (EROS) under the United States Geological Survey. In practice, data use involves careful attention to metadata, which is information about the data that supports interoperability and allows systems to work together. The guidelines developed by European Union to support the interoperability and exchange of geospatial data allow accessibility of data through the Infrastructure for Spatial Information in Europe (INSPIRE). The data restrictions in water resources management are a barrier to researchers as they hinder sharing of knowledge and development of tools.

Different types of water scarcity indicators have been proposed and explained in the literature, such as the water stress or Falkenmark indicator, the water resources vulnerability index, the economic water scarcity index, and the water poverty index, among many others. Interestingly, while regions have to be determined as water secure or insecure according to these indicators, without explicit consideration of environmental factors, the shared conclusion that water security has been achieved by the developed world, such as many countries in Europe, North America, Australia and Japan, is flawed (Rijsberman et al. 2006). Sources of uncertainty can be identified through risk-based water security indicators and can include the uncertainty in the climate model outputs. These indicators are important in risk analysis, which provides essential evidence to inform the choice between alternative courses of action and explores the range of possible future conditions from business-as-usual scenarios to extremely unlikely conditions (Hall and Borgomeo 2013). Hall and Borgomeo (2013) furtherexplained that this decision-making process involves choices between different courses of action based on their benefits in terms of risk reduction and costs. The authors also proposed a framework with a focus on water security risks for the household, industrial and farming sectors. It also presents the complete picture of water security risk developed from the combined set of hazard and exposure indicators, which track contributing factors such as the frequency of weather extremes, declining water quality and other issues related to the assessment of water-related risks.

1.4.1 Associated Factors and Approaches

More than half the global population resides in urban regions, which represent hotspots of locally focused development that place enormous pressure on the surrounding, predominantly freshwater, resources (Podowski and Gorelick 2014).

The continuously increasing demand for water not only alters the hydrological balance of the water cycle, but also puts pressure on the watershed. Podowski and Gorelick (2014), examined the vulnerability of the surface and groundwater supply of 70 cities with populations greater than 7 million and created a reference scenario from the year 2010. Using this reference, the authors formulated a scenario for the year 2040 (Mukheibir et al. 2015; Podowski and Gorelick 2014) based on drivers such as population growth, increasing water demand and agricultural and environmental demands. It was estimated that the vulnerability of 31 of those 70 investigated cities will increase by 20 to 30% by the year 2040, and the cities that are dependent on river/surface water will become more vulnerable and will be dependent on reservoir water supplies. The studies stressed that large cities must be prepared in terms of future management responses to meet the increasing urban water demand (Mukheibir et al. 2015; Podowski and Gorelick 2014).

A conventional method for solving water supply issues is dam construction, which can facilitate the storage and allocation of freshwater, as well as clean energy generation. The cumulative future impact of large hydropower-generating dams on river fragmentation will be substantial (Zarfl et al. 2015), and to better understand it, studies have calculated the river discharge at 3500 dam locations using the HydroSHEDS WaterGAP Global Hydrology Model. The global freshwater WaterGAP model accounts for anthropogenic impacts on the water system such as the construction of dams and the abstraction of ground and surface water, and it calculates the flows and storages of the water system (Van Loon et al. 2016). WaterGap facilitates the determination of historical occurrences of water scarcity in the world's river basins and can be applied to assess water stress and the impact of water-related disasters as well as to quantify the impact of human actions on the current water situation. The study by Zarfl et al. (2015) revealed that hydropower-generating dams partially reduce the energy gap, but they are not able to substantially reduce greenhouse gas (GHG) emissions. In the future, the construction of hydropower dams will affect ecologically sensitive regions such as the Mekong, Amazon and Congo basins, and the authors suggested that it is vital that the boom in hydropower dam construction be reduced to save our planet (Zarfl et al. 2015). Additionally, there is an urgent need to evaluate and mitigate the socioeconomic and environmental ramifications of dam construction globally because, despite its renewable nature, the technology is associated with severe adverse socioeconomic and ecological impacts such as the fragmentation of free-flowing rivers, the relocation of indigenous people, transboundary conflicts and negative impacts on freshwater biodiversity (Zarfl et al. 2015). Climate change is also expected to accelerate global hydrological cycles (Oki and Kanae 2006; IPCC 2014), and its probable impacts on water resources will be seasonal variation, water quality degradation and changes to the hydrological cycle. Precipitation is expected to become more intense, and the risks of floods and droughts will increase (Menon et al. 2002). A fraction of the global population will experience water scarcity, and the other fraction will be affected by floods. In this scenario, the impacts of climate change are projected to reduce the quantity of surface and groundwater resources in most dry subtropical regions and increase the risk of floods and water-related disasters in wet regions, which will globally intensify competition for water (Menon et al. 2002). Studies have shown that water security can be achieved, even in a short time, by addressing the key water governance issues and establishing suitable management strategies (Saraswat et al. 2016), and the new water security paradigms can be transformed to solve issues related to inefficient water use such as non-revenue water (NRW), wise decision for both supply and demand, quality degradation, disaster risk reduction and more. In this respect, the most important problem in achieving water security broadly remains the minimization of inefficient use of water globally, specifically in Asian countries. Table 1.1 shows the water supply service performance indicators for Asian cities, and approximately 25–55% of the drinking water is unaccounted for water (NRW). A major problem lies in water loss due to leaks in the water supply systems in developing economies such as India (NRW of approximately 55%), Indonesia, the Philippines, Sri Lanka and Bangladesh (ranging from 30 to 55%) (Global Water Intelligence 2014). Another key reason is underinvestment in water infrastructure and creating institutions to renovate or rebuild aging infrastructures and manage water resources sustainably. For most of the wasted water due to inefficient use by households as well as in commercial and farming practices, most critical reasons are inefficient pricing policy, subsidizing high-income households and absence of legal penalties for wastage and overuse. In contrast, poor people, who need subsidies, do not have access to piped water systems and often pay proportionately more for the supplied water or must use water from contaminated sources (UNICEF and WHO 2011).

In the current context, the focus has shifted to another very important dimension of water security, which is the treatment of wastewater. Historically, wastewater has always been considered disposable as opposed to a valuable resource. In fact, if treated properly, wastewater can be reused and can contribute to reducing the pollution in water bodies. Water recycling involves reusing treated wastewater or stormwater for many purposes, such as irrigation, flushing toilets, landscape maintenance, industrial needs, and in some cases, groundwater recharge (Saraswat et al. 2016; Kumar et al. 2016; Vigneswaran and Sundaravadivel 2004). Recycled or reused water can provide many types of services and financial savings (Rahman et al. 2014), and if tailored to meet quality standards, it could be a vital resource due to its multiple benefits such as its use for irrigation, cooling power plants and refineries, flushing toilets at the household level, controlling dust in cities, mixing and preparing cement for construction activities, beautifying artificial lakes and parks and for industrial processes. Thus, there are ample reasons to consider wastewater treatment as a sustainable solution for achieving water security. Given an uncertain future, another dimension of the problem that is highly visible in designing sustainable water management strategies and policies is the water resource governance structure. Reports by the IPCC have indicated that more than 87% of the impacts of climate change will be on water-related infrastructure, and the increasingly negative impacts of global warming are likely to increase the variation, frequency and severity of weather such as extreme droughts and floods (Kovats and Akhtar 2008). Currently, many parts of the world suffer from severe flooding and tropical storms throughout the year while others are subject to drought. These weather-related disasters have huge direct and indirect economic impacts (Ding et al. 2011), and they affect agricultural production and can result in the loss of power in urban areas due to insufficient water for hydroelectricity among many other impacts.

There is an urgent need to align our development agendas to achieve the UN SDGs by 2030 especially related to water, and the adopted solutions and processes must aim to achieve water security that addresses a range of development criteria, such as environmental sustainability, economic growth and employment, renewable energy generation and the development of multiple other priorities. Investment in water security is a long-term strategy that will benefit human development and economic growth in the long run with immediate and apparent short-term gains, as well. One of the most significant challenges in operationalizing sustainable water security solutions is the range of probable variables and potential methods, so operationalization at various levels will likely require specific and narrow timeframes to address localized water scarcity issues. The integrative framing of water security needs should occur at the policy level and in governance processes, where priorities should be established and decisions should be made with the agreement of stakeholders. An integrative approach is a likely solution to bring about good water governance and establish new approaches for sustainable water management.

1.4.2 Sustainable Water Management

The complexities of water security solutions ensure that there will be no one-size-fits-all solution for managing water scarcity, so interdisciplinary collaborations are required across sectors, communities and political borders to manage the competition for or conflicts over water resources. The key to achieving water security is sustainable water management, which focuses on the efficient use of water sources, such as precipitation, groundwater and surface water, and on water allocation strategies that increase economic and social returns and enhance water productivity. Furthermore, there is a need for a special focus on equity in access to water as well and the social impacts of water allocation policies. The range of sustainable solutions presented in this chapter will help policy-makers and local water planners understand the challenges of water security issues and deal with local situations more efficiently. In this section, a wide range of sustainable, simple and localized solutions are presented that have the potential to promote greater collaboration in the implementation of strategies to mitigate water-related issues.

Water conservation technologies (WCT) are of greater importance in countries characterized by traditional irrigation schemes with efficiencies of 30–40%, such as India (Loganandhan et al. 2015). Batchelor et al. (2014) explained that the increased use of WCT will play an important role in improving the productivity of rain-based agriculture and irrigation efficiency. Water conservation and water-saving technologies can deliver benefits, including improved water productivity and farm profits if they are well planned and managed. In addition, WCT reduce non-beneficial consumptive water use, such as through well-managed drip irrigation or orchard crops, and enhance the distribution of profits under regulatory frameworks.

Reclaiming treated wastewater is the most economically feasible and sustainable alternative for achieving sustainable water resource management (Leung et al. 2012). Reducing the large amount of water used for flushing toilets in urban areas would reduce water consumption and benefit the implementation of water security strategies. For example, in the Irvine Ranch Water District of California, flushing toilets in high-rise office buildings with reclaimed water began in 1991, and this reuses a significant amount of wastewater (Leung et al. 2012).

Seawater desalination is a universally appropriate option because it produces good-quality potable water, but the cost is very high. To alleviate the expense, the government of Hong Kong implemented a dual water supply system in the 1950s that provides freshwater for potable uses and seawater for flushing toilets. Flushing toilets with seawater and treating the sewage are a much simpler approach than treating seawater to supply freshwater and reuse wastewater. The system reduces freshwater demand, requires less energy consumption and emits fewer greenhouse gases in the process. Researchers have estimated that more than 20% of the municipal water and up to 4% of the total electricity consumption of an urbanized coastal city can be saved by flushing toilets with seawater (Leung et al. 2012; Grant et al. 2012).

The sustainable solution of treatment wetlands has become an attractive choice for removing nutrients from wastewater effluents because of the cheaper operational costs and low-energy requirements as well as the benefits of aesthetically attractive spaces and habitats for flora and fauna (Jasper et al. 2012), and it is hoped that using wetlands to treat a wider range of contaminants will lead to this solution being more extensively applied as a component of urban water infrastructure. Importantly, based on the quality of the treatment, the treated effluents can be reused for various purposes such flushing toilets, cleaning, cooling, as water supply for natural wetlands, and recharge the groundwater in infiltration areas.

Using the landscape may leapfrog the development of urban water services by means of eco-friendly technologies, which are a cheaper, more flexible and adaptable option than the conventional 'gray' technologies that cities in the developing world are currently attempting to implement (Head 2014). Sustainable water management can be achieved by facilitating a swift change in mindset toward landscape-based water management by identifying and engaging likely champions among relevant city officials as well as in flood-prone communities to build a knowledge-sharing process, and it is essential to utilize the pace of urbanization to introduce sustainable construction practices by drafting exemplary water catchment plans in collaboration with stakeholders and communities. Such planning should be based on a physical analysis of existing water management practices and interactive designs, which are necessary for adopting smart distribution technologies and practices for sharing knowledge with communities. Another set of sustainable solutions, as explained by Bazilian et al. (2011), focuses on the critical nexus of water, food and energy that is essential for addressing sustainable development challenges. Water, food and energy are interdependent and share many characteristics; therefore, sustainable solutions should be able to effectively address this interdependency. However, this nexus faces resource constraints due to dependence on healthy ecosystems, sharing of the global good through trade and variations in availability, as well as supply and demand.

Solutions that are based on ecosystem integrity can play an important role in changing the dynamics involved in attaining water security, which remains unexplored territory. The relationship between ecosystems and water security is mutually beneficial as it ensures that sufficient and good-quality freshwater is available to support ecosystem functioning. Ecosystems provide the water required for human, plant and animal communities, so to achieve water security, ecosystems must be conserved because they are vital for sustaining the quality and quantity of the available water within a watershed and provide benefits, such as clean water and drought and flood mitigation, and support the availability of soil water to achieve food security. Conversely, to ensure that ecosystems function properly, basic water security is needed in the form of sufficient, good-quality freshwater. The living things on our planet have been able to adapt to ecosystems because our current hydrological cycle has provided a period of relative stability on which we have come to rely (Sandford, 2012), so altering ecosystem functioning can be disastrous. Although ecosystems can cope with and adapt to variations and significant changes in water quantity and quality, the rapid pace of human development is threatening this capacity and raising concerns that ecosystems will reach a tipping point, past which they will no longer be able to provide services and sustaining functions (Maas 2012).

Further investigation of local adaptation strategies for achieving water security and effectively fighting the impacts of climate change is necessary for policy-makers (Padowski and Gorelick 2014). Current research by highlights that the water and social agreements for a water-sensitive city would significantly differ from conventional approaches to urban water management. The normative values of environmental protection, supply–demand security, flood control, human well-being and economic sustainability should be integrated (Brown et al. 2008).

As a component of sustainable water management, water governance reformation is an attractive and widely accepted concept, but it is more than national-level water legislation, regulations and institutions. Water governance also refers to processes that promote stakeholder or community participation in the design of water and sanitation systems and that empower communities to make decisions about these systems. Restructuring water governance methods requires social mobilization and actions designed to encourage ownership, capacity building, coinvestment, willingness-to-pay for services and incentives for participation at the community level. Water governance solutions build organizational capacity from the local level to the management level, empower stakeholders with knowledge that increases their ability to make decisions and promote gender equality in decision-making (Saraswat and Kumar 2016). Furthermore, they determine the relevant roles for government entities to ensure the delivery of water and sanitation while maintaining checks on public and private players in meeting these needs. The process of building sustainable governance to achieve water security is very effective in correcting market distortions, providing incentives and promoting affordable pricing.

The innovative solution of strengthening the adaptive capacity and water governance of countries can play an important role in eradicating water poverty as tailoring region-specific integrated water resource management to water governance will serve local needs and purposes. Good water governance is a continuously evolving process that requires enhancement as it responds to new problems, experiences and challenges (Rogers and Hall 2003), and it is based on empowered institutions, well-designed and effective legislation and strong policy instruments to achieve the social, economic and environmental goals associated with water security. For example, Japan's IWRM promotes sustainable water use through a policy framework appropriate for achieving a sound water cycle. Tailored IWRM combines the perspectives of surface and groundwater, water supply and demand, upstream and downstream water quality and the need to promote water use efficiency and environmental conservation. The critical issues in establishing good governance are strong coordination and cooperation between various departments within a country as well as across borders, developing policies and incentives to draw people away from unsustainable groundwater extraction, reconciling water rights and empowering stakeholders with accurate information to participate in decision-making.

1.5 Summary

Water security is a web that links food and energy security, climate, economic growth and human well-being. Water is vitally important to ecosystems and human societies, but the combination of urbanization, population growth, socioeconomic change, evolving energy needs and climate change are exerting unprecedented pressure on the utilization of freshwater resources. Water security involves ensuring that every person has reliable access to sufficient safe water at an affordable price to enable a healthy and productive life, as well as maintaining water-related ecological systems for future generations. The problems associated with water conflicts are well-recognized, and the United Nations Sustainable Development Goals will not be achieved without solving problems related to water security. Although the available water resources are sufficient to satisfy the total water demand at the global scale, shortages prevail across different spatial and temporal scales. Constraints on water availability and deteriorating water quality threaten secure access to water resources for different uses, and despite recent progress in developing new strategies, practices and technologies for water resource management, as well as their dissemination and implementation, have been limited. A comprehensive sustainable approach to addressing water security challenges requires connecting social, economic and environmental systems at multiple scales.

This chapter explored the new paradigm and evolving definition of water security to solve the water crisis. Through case studies, localized solutions to water problems that integrate locally available methods were shown to be sustainable and effective, which can lead to the overall water security of a region. In most cases, the most efficient strategy to cope with and adapt to the impacts of water scarcity was capacity building to predict problems and respond in a timely way. Providing humanitarian actors with access to updated information about water-related hazards is crucial for adaptation to water scarcity challenges as is ensuring that information is correctly interpreted and applied in policy design. It is an issue that most of the available climate- or water-related information that currently exists is too technical and thus not easily understandable to local policy- or decision-makers, so there is an urgent need to translate this information for practical applications. Therefore, improving the use of climate information is essential, and it is important to convey information about water-related disasters in a simple and understandable way.

It is widely accepted that water security is not only the ability to provide enough water to the inhabitants of a region. The evolving paradigm of achieving water security also includes providing sufficient water for domestic, industrial and commercial socioeconomic activities as well as clean drinking water to meet basic needs at an affordable price with proper sanitation to ensure human well-being. Furthermore, it includes the treatment and collection of wastewater to curb pollution. However, achieving water security also extends to protecting the ecosystem when extracting water resources and increasing the role of natural infrastructure to sustain ecosystem functioning as well as developing the ability to cope with and mount a timely response to water-related disasters such as droughts, floods, disease spread and pollution. To ensure the sustainability of water resources, transboundary water resource management approaches must be implemented, and good water governance that includes planned, well-operated, transparent and accountable departments to develop capacity and maintain the water infrastructure is also needed to achieve water security.

This makes water security a complex problem that varies according with local and regional contexts. Both biophysical extremes, such as floods and drought, and social extremes, such as gender equality, are critical to the process of formulating water management policy, but another important link is to the resources such as land-use change and food security. As highlighted in this chapter, the need for a reliable and accessible evidence base for water management is inescapable, so there is a need for a huge investment in data collection and the development of models and tools to promote integration and visualization. There is also a need for open, web-based access to data through national and international cooperation. To attain water security in a changing context, there is not a simple, single or one-size-fits-all solution, but it is recognized that there is a need for a portfolio of solutions that are appropriate for varying contexts, consider local societal values, leverage multiple water uses and are synergistic with the management of other resources. The shift from linear/single pass through water use to the circular use of water and the embedded resources, such as nutrients and energy, is an example to facilitate the understanding of the changing water security context. Expanding the solution portfolio from fully engineered systems to managed systems including capacity building, community awareness, wetland management and aquifer conservation is needed to incorporate natural processes to achieve a water-secure future for all and sound economic development for human well-being. The challenges of attaining a water-secure, sustainable future are substantial, but there are also many case studies of good practices using various measures from technological advancements to natural infrastructure, and there is a need to promote and expand successful approaches. This chapter presented a wide range of sustainable solutions for achieving water security in a changing context and explored and analyzed the application of sustainable water solutions around the world to understand the best practices in the field and their applicability in local contexts. This is the time to make decisions to achieve water security and a sustainable future by providing the best possible evidence base and improving communication with local and national water planners and decision- and policy-makers. Past solutions will not be sufficient to meet future challenges; therefore, there is an urgent need for research into new alternatives and education that addresses the need for more integrated solutions.

On policy recommendation side, a holistic approach and long-term planning will be required to achieve water security. Many studies have indicated that significant progress has been achieved in several areas, but the solutions still lack social, economic and environmental dimensions and undermine the importance of localized approaches. Several challenges, such as the gap between water demand and supply, old and aging water infrastructure, unreliable sanitation services, wastewater treatment and collection, declining water quality, lack of preparedness for water-related disasters and ecosystem degradation, undermine human well-being and are a danger to the planet. To address these issues and improve water security, tailored policy responses are needed that can be adapted at local, regional, national and international levels. As illustrated by the case studies presented above, countries with tailored and flexible policies and foundations as well as strong legal frameworks can efficiently adapt to the impacts of water stress. In a changing context, the evolving definition of water security requires a wide range of sustainable solutions, and in many places, approaches that mix structural and non-structural measures with strong governance are required to address water scarcity.

References

ADB annual report (2008) Managing Asian Cities. Manila

Asano T (2002) Water from (waste) water–the dependable water resource (The 2001 Stockholm Water Prize Laureate Lecture). Water Sci Technol 45(8):23–33

Asian water development outlook (AWDO) (2016) Strengthening water security in Asia and the Pacific. (n.d.). Retrieved from https://www.adb.org/publications/asian-water-development-outlook-2016

Batchelor C, Reddy VR, Linstead C, Dhar M, Roy S, May R (2014) Do water-saving technologies improve environmental flows? J Hydrol 518:140–149

Bazilian M, Rogner H, Hoells M, Hermann S, Arent D, Gielen D, Steduto P, Mueller A, Komor P, Tol RSJ, Yumkella KK (2011) Considering the energy, water and food nexus: towards an integrated modelling approach. Energy Policy 39:7896–7906

Brown R, Keath N, Wong T (2008) Transitioning to water sensitive cities: historical, current and future transition states, paper submitted to 11th international conference on urban drainage, Edinburgh, Scotland, UK

Brown RR, Keath N, Wong T (2009) Transitioning to water sensitive cities: historical, current and future transition states. En 11th International Conference on Urban Drainage. Conferencia presentada en Edinburgh, Scotland, UK from 31st August to 5th September, 2008, IWA publishing house, pp 10

Cashman A (2014) Water security and services in the Caribbean. Water 6(5):1187–1203

Chellaney B (2011) Water: Asia's new battleground. Georgetown University Press
Cohen B (2006) Urbanization in developing countries: Current trends, future projections, and key challenges for sustainability. Technol Soc 28(1):63–80
Conca K (2006) The new face of water conflict. Woodrow Wilson International Center for Scholars vol 3
Connor R (2015) The United Nations world water development report 2015: water for a sustainable world. UNESCO Publishing vol 1
Cook C, Bakker K (2012) Water security: debating an emerging paradigm. Glob Environ Change 22(1):94–102
Ding Y, Hayes MJ, Widhalm M (2011) Measuring economic impacts of drought: a review and discussion. Disaster Prevent Manage 20(4):434–446. https://doi.org/10.1108/09653561111161752
Falkenmark (1989) The massive water scarcity threatening Africa-why isn't it being addressed. Ambio 18(2):112–118
Gareau BJ, Crow B (2006) Ken Conca, Governing water: contentious transnational politics and global institution building. Int Environ Agreements Politics, Law Econ 6(3):317–320
Garrick D, Hall JW (2014) Water security and society: risks, metrics and pathways. Annu Rev Environ Res 39:611–639
Gleick PH, Ajami N (2014) The world's water volume 8: the biennial report on freshwater resources. Island press. vol 8
Global Water Intelligence (2014) Market Profile: Smart water networks. Global water intelligence, January 2014, pp 39–41
Global Water Intelligence (2016) Global Water Market 2017: meeting the world's water and wastewater needs until 2020. CWC Publications. pp 1832
Global Water Partnership (GWP) (2015) Promoting effective water management cooperation among riparian nations. The background paper, number 21. Stockholm, Sweden, p. 64
Grant GB, Saphores JD, Feldman DL, Hamilton AJ, Fletcher TD, Cook PLM, Stewardson M, Sanders BF, Levin LA, Ambrose RF, Deletic A, Brown R, Jiang SC, Rosso D, Cooper WJ, Marusic I (2012) Taking the "waste" out of "wastewater" for human water security and ecosystem sustainability. Science 337:681–686
Grey D, Garrick D (2012). Water security as a 21st century challenge. Brief No.1. In: International conference on water security, risk and society, University of Oxford, 16–18 April
Grey D, Sadoff CW (2007) Sink or swim? Water security for growth and development. Water Policy 9:545–571
Hall JW, Borgomeo E (2013) Risk-based principles for defining and managing water security risk-based principles for defining and managing water security Phil. Trans R Soc A 371:0407
Hanjra MA, Qureshi ME (2010) Global water crisis and future food security in an era of climate change. Food Policy 35(5):365–377
Head BW (2014) Managing urban water crises: adaptive policy responses to drought and flood in Southeast Queensland, Australia. Ecol Soc 19(2):33
Henze M, Harremoes P, la Cour Jansen J, Arvin E (2001) Wastewater treatment: biological and chemical processes. Springer Science & Business Media
Hering JG (2015) Do we need more research or better implementation through knowledge brokering? Sustain Sci. https://doi.org/10.1007/s11625-015-0314-8
Huitema D, Mostert E, Egas W, Moellenkamp S, Pahl-Wostl C, Yalcin R (2009) Adaptive water governance: assessing the institutional prescriptions of adaptive (co-) management from a governance perspective and defining a research agenda. Ecol Soc 14(1):26
Human Development Report (2006) UNDP, 2006
IPCC (2007) Climate Change 2007: The physical science basis. Contribution of working group I to the fourth assessment report of the intergovernmental panel on climate change
IPCC (2014) Climate change 2014: synthesis report. In: Pachauri RK, Meyer LA (eds.), Core writing team, contribution of working Groups I, II and III to the fifth assessment report of the intergovernmental panel on climate change. IPCC, Geneva, Switzerland, p. 151

Jain R (2012) Providing safe drinking water: a challenge for humanity. Clean Technol Environ Policy 14(1):1–4

Japan International Cooperation Agency (JICA) (2014) Annual report, pp 180. https://www.jica.go.jp/english/publications/reports/annual/2014/c8h0vm000090s8nnatt/2014_all.pdf. Accessed on 15 March 2017

Jarraud M (2015) UN Water Annual report. pp 32

Jasper C, Le TT, Bartram J (2012) Water and sanitation in schools: a systematic review of the health and educational outcomes. Int J Environ Res Public Health 9(8):2772–2787

Jensen O (2016) Public–private partnerships for water in Asia: a review of two decades of experience. Int J Water Res Dev 1–27

Jiang Y (2015) China's water security: current status, emerging challenges and future prospects. Environ Sci Policy 54:106–125

Kovats S, Akhtar R (2008) Climate, climate change and human health in Asian Cities. Environ Urban 20(1):165–175

Kumar P, Kumar A, Singh CK, Saraswat C, Avtar R, Ramanathan AL, Herath S (2016) Hydrogeochemical Evolution and Appraisal of Groundwater Quality in Panna District, Central India. Exposure Health 8(1):19–30

Lankford B (2013) A synthesis chapter: the incodys water security model. In: Water security: principles, perspectives and practices. Earthscan Publications. London. ISBN 978-0-415-53471-0

Lankford B, Bakker K, Zeitoun M, Conway D (2013) Water security: principles, perspectives and practices, Edited, Routledge, pp 376

Leung RWK (2012) Integration of seawater and grey water reuse to maximize alternative water resource for coastal areas: the case of the Hong Kong International Airport. Water Sci Technol 65(3):410–417

Leung HW, Minh TB, Murphy MB, Lam JCW, So MK, Martin M et al (2012) Distribution, fate and risk assessment of antibiotics in sewage treatment plants in Hong Kong, South China. Environ Int 42:1–9

Loganandhan N, Patil SL, Srivastava SK, Ramesha MN (2015) Post-adaptation behaviour of farmers towards soil and water conservation technologies of watershed management in India. Indian Res J Extension Educ 15(1):40–45

Maass JM (2012) El manejo sustentable de socioecosistemas. Pages 89–99 in J. L. Calva, coordinator. Cambio climático y políticas de desarrollo sustentable. Tomo 14 de la colección Análisis Estratégico para el Desarrollo. Juan Pablos Editor- Consejo Nacional de Universitarios, Coyoacán, México

Majumder M (2015) Impact of urbanization on water shortage in face of climatic aberrations. Springer briefs in water science and technology. Springer publications, p 98. ISSN 2194-7244

Malve O, Tattari S, Riihimäki J, Jaakkola E, Voß A, Williams R, Bärlund I (2012) Estimation of diffuse pollution loads in Europe for continental-scale modelling of loads and in-stream river water quality. Hydrol Proc 26(16):2385–2394

Menon S, Hansen J, Nazarenko Luo Y (2002) Climate effects of black carbon aerosols in China and India. Science 27:2250–2253

Mukheibir P, Howe C, Gallet D (2015) Institutional Issues for Integrated' One Water' Management, Water Environment Research Foundation (WERF) in Partnership with the Water Research Foundation (WaterRF) and Water Quality Research Australia (WQRA), IWA Publishing

OECD (2013) Water and Climate Change Adaptation: Policies to Navigate Uncharted Waters, OECD Studies on Water, OECD Publishing. http://dx.doi.org/10.1787/9789264200449-en

Oki T, Kanae S (2006) Global hydrological cycles and world water resources. Science 313(5790):1068–1072

Padowski JC, Gorelick SM (2014) Global analysis of urban surface water supply vulnerability. Environ Res Lett 9(10):104004

Perrot-Maitre D, Davis P (2001) Case studies of markets and innovative financial mechanisms for water services. Forest Trends and the Katoomba Group, Washington, D.C.

Pink RM (2016) Introduction. In: Water rights in Southeast Asia and India. Palgrave Macmillan US, pp 1–14

Prokurat S (2015) Drought and water shortage in Asia as a threat and economic problem. J Modern Sci Tom 235–250

Rahman MM, Hagare D, Maheshwari B (2014) Framework to assess sources controlling soil salinity resulting from irrigation using recycled water: an application of Bayesian Belief Network. J Clean Prod 105:406–419

Rijsberman FR (2006) Water scarcity: Fact or Fiction? Agric Water Manag 80(5):22

Rogers P, Hall AW (2003) Effective water governance. Global water partnership, vol 7

Sachs JD (2012) From millennium development goals to sustainable development goals. The Lancet 379(9832):2206–2211

Sandford RW (2012) Cold matters: the state and fate of Canada's fresh water. Rocky Mountain Books, Victoria. Springer Publications, pp 230

Saraswat C, Kumar P (2016) Climate justice in lieu of climate change: a sustainable approach to respond to the climate change injustice and an awakening of the environmental movement. Energy Ecol Environ 1(2):67–74

Saraswat C, Kumar P, Mishra BK (2016) Assessment of stormwater run-off management practices and governance under climate change and urbanization: an analysis of Bangkok, Hanoi and Tokyo. Environ Sci Policy 64:101–117

Sergiusz P (2015) Drought and water shortages in Asia as a threat and economic problem. Journal of modern science 3(26):235–250

Tremblay H (2010) A clash of paradigms in the water sector? Tensions and synergies between integrated water resources management and the human rights-based approach to development. Tensions and synergies between integrated water resources management and the human rights-based approach to development, 1 Aug 2010

UNDESA (United Nations Development of Economic and Social Affairs) (2016) International Decade for Action, WAter for Life' 2005–2015. URL: http://www.un.org/waterforlifedecade/scarcity.shtml

UNEP (2016) A Snapshot of the World's Water Quality: Towards a global assessment. United Nations Environment Programme, Nairobi, Kenya. pp 162

UNICEF and WHO (2011) Drinking water: equity, safety and sustainability. http://www.wssinfo.org/fileadmin/user_upload/resources/report_wash_low.pdf

United Nations, Department of Economic and Social Affairs, Population Division (2014). World Urbanization Prospects: The 2014 Revision, Highlights (ST/ESA/SER.A/352)

United Nations University (UNU) (2013) Water security and the global water agenda. Research brief. UNU-INWEH publisher, Hamilton, Canada, p. 38

UN-Water (2010) UN-Water annual report 2010, Geneva, Switzerland, pp 21. Available at: http://www.unwater.org/publications/UN-Water_Annual_Report_2010.pdf

UN-Water (2013) UN-Water annual report 2013, Geneva, Switzerland, pp 40. Available at: http://www.unwater.org/publications/UN-Water_Annual_Report_2013.pdf

UN-Water, FAO (2007) Coping with water scarcity. Challenge of the twenty-first century

Van Loon AF, Gleeson T, Clark J, Van Dijk AI, Stahl K, Hannaford J, Di Baldassarre G, Teuling AJ, Tallaksen LM, Uijlenhoet R, Hannah DM (2016) Drought in the Anthropocene. Nat Geosci 9(2):89–91

Vigneswaran S, Sundaravadivel M (2004) Recycle and reuse of domestic wastewater, in wastewater recycle, reuse and reclamation. In: Vigneswaran S (ed.) Encyclopedia of Life Support Systems (EOLSS), Developed under the Auspices of the UNESCO, Eolss Publishers, Oxford, UK. http://www.eolss.net

Vörösmarty CJ, McIntyre PB et al (2010) Global threats to human water security and river biodiversity. Nature 467:555–561. https://doi.org/10.1038/nature09440

Wagener T, Sivapalan M, Troch PA, McGlynn BL, Harman CJ, Gupta HV, Kumar P, Rao PSC, Basu NB, Wilson JS (2010) The future of hydrology: an evolving science for a changing world. Water Resources Res 46(5)

Winter TC, Harvey JW, Franke OL, Alley WM (1998) Ground water and surface water: a single resource. U.S. Geological Survey Circular report, Report number 1139. USGS Publication Warehouse. New York, USA. pp 89

World Resources Institute (2015) Aqueduct Projected Water Stress Country Rankings by Luo T, Young R, Reig P. Technical Note. Washington, D.C., World Resources Institute, pp. 16. Available online at: www.wri.org/publication/aqueduct-projected-water-stresscountry-rankings

World Wildlife Fund for Nature (WWF) (2012) Technical report on water. https://wwf.panda.org/knowledge_hub/all_publications/living_planet_report_timeline/lpr_2012/demands_on_our_planet/water_footprint/. Accessed on 16 August 2018

WSSD (2002) In Lankford B, Bakker K, Zeitoun M, Conway S (eds.). Water security: principles, perspectives and practices (2013). NY, Routledge

Zarfl C, Lumsdon AE, Berlekamp J, Tydecks L, Tockner K (2015) A global boom in hydropower dam construction. Aquat Sci 77(1):161–170

Chapter 2
Urban Water Security Challenges

2.1 Background

Water is essential for the survival of animal and plant life. At global scale, availability of usable water resources is well above total demand (Oki and Kanae 2006). However, the distribution of water resources is not uniform either spatially or temporarily. There is a definite surplus of water during the rainy season while significant water deficit occurs during dry season. For increasing water security in urban areas, the phenomena like devastating floods, reliable water supply for drinking, irrigation, hydropower among others that are associated with excessive and scarce water must be managed well. The rapid growth of population, with extension of irrigated agriculture and industrial development has increased the demand of water. As the Earth's population has been growing rapidly, more stress is being put on the land to support the increased population. One question that arises is how water resources will be affected. In several cases, water shortages have led to conflicts over the rights of the water. Environmental issues such as provision of clean water, production and processing of wastewater and flooding and land subsidence are frequently reported in many cities. With industrial and domestic water demand expected to double by 2050, competition among urban, peri-urban and rural areas will further deteriorate (Jalilov et al. 2018; UNDP 2006). In many urban areas, the sustainable use of water is approaching or exceeding the limits (Hatt et al. 2004; Mitchell et al. 2003). In the urban basin, competition between agriculture and industry is intensifying (Bahri 2012). The unplanned urbanization is highlighted by the degraded environment in the urban areas due to changes in hydrology of catchments and modified riparian ecosystems.

In addition to various influences, it is important to consider the effects of two crucial issues: urbanization and climate change on water environment, such as surface runoff, stream flow, flooding, river pollution, biodiversity, and groundwater recharge. In context of climate change and urbanization, the negative impacts are exacerbating resulting greater runoff, pollutant loads and pressure on existing systems.

B. K. Mishra et al., *Sustainable Solutions for Urban Water Security*,
Water Science and Technology Library 93,
https://doi.org/10.1007/978-3-030-53110-2_2

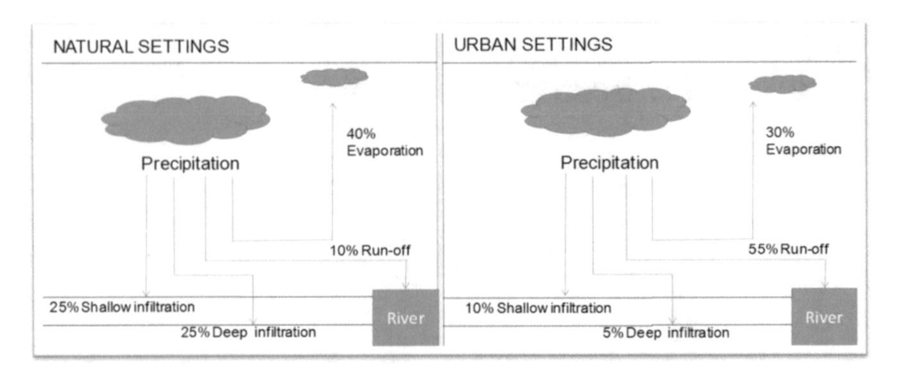

Fig. 2.1 Illustrative sketch for the impacts of land-use change on water cycle (Adopted from Saraswat et al. 2016)

Urbanization leads to the formation of an impermeable surface, which increases the runoff and downstream floods and less recharge of groundwater (Fig. 2.1). Ultimately, the loss of recharge affects the supply of residential and municipal water. As industrialization continues, more of the world's population becomes concentrated in urban areas, with greater stress on available water resources in smaller area. Builtup areas have increased through both formal as well as nonformal businesses and settlements (Kefi et. al. 2018). This increase has not been adequately supported by the increases of environmental service capacities to check the consequence of the development.

Similar to urbanization (land-use change), climate change affects local hydrological cycles by producing more surface runoff and decreasing the base flow, interflow and depression storage. Community planners, developers and citizens should be aware of the impacts of land-use and climate change on their environmental resources. Anticipating the rate, amount and duration of rainfall in each heavy rain event under climate change is highly important in planning and designing of stormwater management facilities, erosion and sediment control structures, flood protection structures and many other civil engineering structures involving hydrologic flows.

It is well accepted that traditional urban water management approach is largely unsuitable to address current and future sustainability issues (Ashley et al. 2005; Wong and Brown 2008). The conventional approach to manage urban water systems, around the world, has seen the use of a similar series of systems for drainage of stormwater, potable water and sewerage. United Nations (UN Agenda 21 1992) defined achieving sustainable urban water systems and protecting the quality and quantity of freshwater resources as key components of ecologically sustainable development. It is important to plan, design and manage water resources system very carefully and intelligently. Minimizing the disturbance on an urbanizing watershed is one way of ensuring continued water supply and decreasing urban floods. Because each land use has different levels of influence, careful physical planning can minimize these effects. Although the influence of urban sprawl on groundwater recharge, and the quantity and quality of surface water is of considerable importance, many planners, urban managers and water resource experts lack to incorporate the potential hydrological impacts of climate and land-use change. Science-based system that justifies the relative impact of both urbanization and climate change on stormwater runoff at local scale is largely warranted in the changing environment. This chapter focuses on the risks of two major threats: urbanization and climate change on water security which is defined as 'the reliable availability of an acceptable quantity and quality of water for health, livelihoods and production, coupled with an acceptable level of water-related risks.' Water security also means addressing the adverse effects of environmental protection and poor management. It is also concerned with ending fragmented responsibility for water and integrating water resources management across all sectors-finance, planning, agriculture, energy, tourism, industry, education and health. A water-secure world reduces poverty, advances education and improves living standards.

2.2 Urbanization

The urban population of the world has grown rapidly from 746 million in 1950 to 3.9 billion in 2014 (UN-DESA 2014). Since the Universal Declaration of human rights in 1948, cities and towns are believed to be a norm of primary human living standards and are now globally accepted because of the many benefits city space provides to the improvement of human well-being. According to the World Economic and Social Survey 2013 report, a net 1.3 billion people were added to small urban cities, 632 million added to medium and 570 million was added to the large urban centers, in total about 2.5 billion people were added to the global urban population between 1950 and 2010. The United Nations Population Division marked 2007 as a year when the number of people living in urban areas became more than half of the world total population (Fig. 2.2). The global urban population has further been predicted to exceed 70% by 2050, estimated to be 6.25 billion, and about 80% of this figure has been projected to live in developing nations and highly concentrated in Africa and

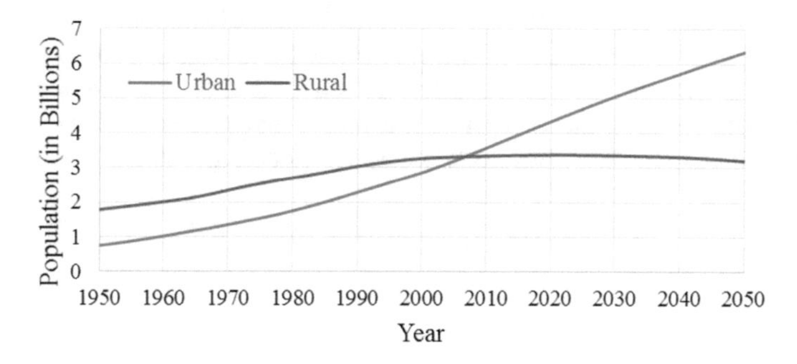

Fig. 2.2 Annual urban and rural population estimates of the world during 1950–2050 (*Source* 'World Urbanization Prospects: The 2014 Revision' database)

Asia. Ismail (2014) recently reported that, at regional level over the past decades, the southeast Asian region has seen greater increase (nearly two times during 1985–2010) in the population in urban area. Unprecedented increase in urban population has added several challenges such as slum expansion and deterioration of water environment.

Rapid urbanization and economic growth have resulted in widespread environmental degradation in urban areas. The imbalances are unavoidable, because urbanization leads to significant demographic, sociocultural, environment and political changes that eventually affect the realization of the idea of urban sustainability. City is the center of economic growth, job creation, innovation and cultural exchange. This is because the cities in developing countries are largely concentrated on production activities providing greater opportunity for additional income. Cities provide better woman participation, health access, literacy rate and upward social mobility. Grubler and Fisk (2013) found that the average urban gross domestic product (GDP) accounts for about 80% of world GDP. The concentration of people, economic activities and services in relatively small areas has big impact on urban society and the economy. Urban economies lead to better access to services, higher prosperity and lifestyle changes, but rapid urbanization also leads to increased slums and squatter settlements, social alienation and environmental pollution. Positive and negative impacts of urbanization are not evenly distributed among urban populations. Rich and powerful are more benefited from positive effects and better protected from adverse effects than poor and alienated people.

In developing countries, urbanization is occurring at a high rate. In 2013, the United Nations Economic and Social Division reported that the land-use patterns and urbanization is diverse in developed and developing countries. The growth of cities in developing countries is at rapid pace and often concentrated in the capital. Urbanization leads to increased impervious surface areas, and the construction of stormwater drainage networks increases the time of concentration and direct runoff, thereby resulting in a more rapid rise in flow rate and depletion of groundwater table (Saraswat et al. 2016, 2017; Willems et al. 2012). In addition, natural water bodies such as lakes, wetlands which can hold a significant amount of flood water are being reduced or filled resulting increased incidence of flooding. The city's growth, reported by Seto et al. (2011), drives the change of local regional environment by creating the most human-dominated landscape and transforming the land cover, hydrology system and organisms. The hiccups are that these areas do not have the capacity to invest in building resilience to these possible disasters. Therefore, these require innovative, locally focused, new analytical space modeling methods that include all variables of land-use changes.

2.3 Climate Change

Water management is planned based on local conditions. Climate change has major impact on the water resource system (Mishra et al. 2019a, b; Kharrazi et al. 2017; IPCC 2014). Recognizing the importance of water resources, many studies have been done to investigate the effects of climate change on precipitation patterns and hydrological structures. These studies suggested varying regional trends (increasing or decreasing) in the future due to climate change. In particular, precipitation increases in wet areas, decreases in arid areas. Most of the studies indicated that heavy precipitation events (frequency and intensity) will increase in future. Changing weather regime will bring prolonged droughts and excessive rains.

Climate change impacts such as the amount, timing and intensity of rain events, in combination with land development, can significantly affect the amount of stormwater runoff that needs to be managed. In some regions, the combination of climate and land-use change may make existing stormwater-related flooding worse, while other areas may be minimally affected. Intergovernmental Panel on Climate Change (IPCC 2014) reported that among the most challenging anthropogenic environmental, economic and social global issue today is climate change. It is reported that climate change is mainly caused by the increase in the concentration of greenhouse GHG in the atmosphere. The carbon emissions are typically attributed to anthropogenic aspects such as land-use changes including agricultural, forestry sectors and city energy demand through fossil fuels and biomass consumption.

Climate change projections are widely used to assess likely future impacts. Global climate models are currently the most credible tools available for simulating the response of the global climate system to increasing greenhouse gas concentrations and provide climatic variables, such as temperature and precipitation. Several GCM projections are available based on some possible scenarios of global warming and CO_2 generation rates. These projections are available for current and future climate (IPCC 2014). Multiple GCMs and scenarios are used to reflect the uncertainty associated climate change. In the climate modeling community, projections are available in terms of four emission scenarios: one mitigation scenario (RCP2.6), two medium stabilization scenarios (RCP4.5/RCP6) and one very high baseline emission scenario (RCP8.5).

Due to great amount of uncertainty associated with the scenarios and projections, use of multiple GCM is recommended to provide the range of recommendations for addressing various climate change impacts. GCM outputs are largely biased when compared with observation data due to flaws in model structure and coarse resolution input. Therefore, direct use of GCM precipitation outputs is considered not suitable for the climate change impact assessment at basin level. Downscaling enables minimization of biases in GCM outputs to be used at local scale climate change impact assessment. There are several downscaling techniques available to transform GCM outputs to local scale for reliable impact assessment. Dynamical downscaling technique converts GCM outputs into local climate data by enhancing atmospheric circulations and climate variables to finer spatial scales using regional climate models. Statistical downscaling techniques use models of correspondence between GCM contemporary climate scenarios data and real-world data.

2.4 Implications on Water Security

Population increase coupled with urbanization and changing climate out constrains water security in many ways. Rapid urbanization and global climate change will greatly alter water environment in developing cities. This section examines both qualitative and quantitative aspects of water security. It identifies water security in the megacity in terms of threats to the environment such as flood relating it to urbanization and climate change. It also examines water security threats in megacity in the case of inadequate water supply relating it to urbanization and climate change.

2.4.1 Hydrological Cycle

In a natural environment, the small percentage of precipitation becomes surface runoff, but as the urbanization is growing and the development expanding, the percentage of stormwater increased abruptly (Fig. 2.1). The surface water runoff is created when pervious or impervious surfaces are saturated from precipitation or snow melt (Durrans 2003). Pervious surface areas absorb the water naturally to the saturation point and after which the amount of the rainwater runs off and travels via gravity to the nearest stream. This point of saturation is dependent on the landscape, soil type, evapotranspiration and the biodiversity of the area (Pierpont 2008).

Greater imperviousness (roads, roofs, pavement) resulting increased surface runoff, reduced infiltration and less groundwater. Increased drainage network also results less ground recharge and quick peak discharge. As it is well-known fact that the urbanization alters watershed hydrology as land becomes more and more covered with surfaces impervious to rain, water is redirected from groundwater recharge. In a natural setting, only very small percentage of precipitation becomes surface runoff, but as the urbanization is growing and the development expanding, this percentage of stormwater increases abruptly. This runoff normally flows into the nearest stream or river and increases the percentage of water in the system, and if it is polluted, it can lead to disastrous situations. The Center for Watershed Protection has reported that areas that exceed 10% imperviousness, stream health begin to decline (Coffman and France 2002). The urbanized watershed faces increased flooding, stream bank erosion and pollutant export. The receiving streams of these intensified storm flows alter hydraulic characteristics due to peak discharges several times higher than predevelopment. With urbanization, environmental issues such as flooding, solid waste problems, proliferation of informal settlements and air pollution are prevalent in the region. Encroachment (flood plains, obstruction of water/floodways, loss of natural flood storage) results in increased flooding. Green areas hold an important role in maintaining balance in urban environment, because of their main function as water retention slow release, while enhancing inflilration in a catchment.

2.4.2 *Water Shortage*

Access to clean and adequate water remains a critical challenge and an acute seasonal problem in urban areas. Freshwater is crucial to human society—not just for drinking, but also for farming, washing and many other activities. It is expected to become increasingly scarce in the future, and this is partly due to climate change. Rainfall distribution/pattern in a year will alter significantly with the climate change, although it is projected to alter not much at annual scale. Greater rainfall due to climate change leads to more rapid movement of water from the atmosphere back to the oceans, reducing our ability to store and use it. The overall effect is an intensification of the water cycle that causes more extreme floods and droughts.

2.4.3 *Land Subsidence*

Highly urbanized cities rely heavily on groundwater for water supply resulting in uncontrolled withdrawal from groundwater aquifers. Rapid urbanization has reduced aquifer recharge and has resulted in declining groundwater levels as well as salt intrusion and land subsidence. Overextraction of groundwater is now a pressing problem in rapidly growing cities of developing nations. Illegal/unregulated construction of wells has proliferated in the region. In the urban environment, the impervious surfaces which cover the natural environment over the ground, the pattern of hydrological process of surface water runoff becomes more unnatural, causing damage to infrastructure and the impairment of receiving waters by pollutants. Indiscriminate land-use practices has impacted on the quality of surface water and modified the hydrologic conditions (Masago et al. 2019).

In urban areas, land subsidence is mainly caused by excessive groundwater extraction, higher load of constructions and infrastructures, natural consolidation of alluvium soil and natural event such as tectonic activity. For example, land subsidence in Jakarta occurs mainly because of lavish groundwater extraction and higher load of construction and infrastructures. Through years, water demand in Jakarta is gradually increasing and this phenomenon is not supported by an adequate water supply. In the long term, land subsidence would be a potential cause of flood in Jakarta. In fact, land subsidence has proven to be one of the main causes of flood in Jakarta, particularly in North Jakarta. Continuous land subsidence will also endanger drainage structures in Jakarta which can make the flood even worst.

2.4.4 *Surface and Groundwater Pollution*

Demands from intensive development and utilization activities, population explosion, poor environmental management—all these contribute to the poor quality of water in almost all water bodies in urban areas. The unregulated discharge of domestic and industrial wastewater and agricultural runoff had caused extensive pollution of receiving water bodies. The effluent being discharged comes in the form of raw sewage, detergents, fertilizers, heavy metals, chemical products, oils and even solid waste. Each of these pollutants has different noxious effects that influence human livelihoods and translate into economic costs. Pollutants accumulate on impervious surfaces, and the increased runoff with urbanization is a leading cause of nonpoint source pollution. Sediment, chemicals, bacteria, viruses and other pollutants are carried into receiving water bodies, resulting in degraded water quality.

It is estimated that large portion of the total garbage generated are not collected by the solid waste management agencies. The uncollected garbage goes into the river systems resulting in the clogging of waterways. This aggravates flooding in the metropolis. Wastewater is another water management issue that needs to be considered more thoroughly. Higher population density and their activity not only affect the water supply and demand but also the production of wastewater released to the river or open channel. The loss of aesthetic value of rivers and other water bodies is a direct result of the pollution of these water bodies. The presence of informal settlers living along the rivers and their tributaries also contributed to the constriction of the drainage areas, resulting to flooding during heavy rains. These informal settlers also add to the deterioration of the water quality of these water bodies.

2.4.5 *Human Health*

Over the last 2–3 decades, most of the water bodies in Manila have been increasingly under threat of unprecedented and uncontrolled urbanization, industrialization, population explosion in urban centers caused by mass migration into cities, unsound land-use and solid waste management practices, and unabated pollution of water and the air. With the changing climate regime as manifested by the increase in the number of typhoons resulting to flooding, the need to address the risks from a deteriorating environment has become one of the biggest challenge of the modern time. Impaired water quality will endanger community's health, leading to the increasing of poverty level. People with lower economic condition tend to face difficulty in providing their wastewater treatment and drinking water facilities. Slum areas in Jakarta are more prone to diseases from bad sanitation and contaminated water source. People who live in those areas often suffer from diarrhea and intestinal worm infection, mainly during rainy season when flood has worsen the water contamination. The social impact apparently not only threatened community's health but also affected community's capacity and poverty in Jakarta.

The deteriorating quality of the water in major water bodies and urban environment situation in the metropolis is negatively affecting the overall health conditions of the population (Masago et al. 2019). A recent joint Japan International Cooperation Agency (JICA) and NSO Study (January 2011) indicated that there are four main causes of water-related diseases in the Philippines: drinking polluted water, contact with polluted water, infection by vector and infection by parasite. Cholera, typhoid, para-typhoid, hepatitis, dysentery and diarrhea are typical cases resulting from taking in polluted water. Scabies, conjunctivitis, typhus and trachoma are the common diseases that can be contracted from contact with polluted water. Infection by a vector transmits diseases such as malaria, dengue and yellow fever while infection by parasite can give rise to such illnesses as filariasis and schistosomiasis. In Metro Manila, diarrhea is the second or third leading causes of morbidity based on a 5-year average from 1996 to 2000 as well as in 2001.

2.4.6 Ecosystem and Biodiversity

The pressures that come with rapid development and urbanization have put so much stress on the water and the environment resulting in destabilization of ecosystems, destruction of natural habitats and an alarming rate of biodiversity losses. During the last 10–15 years witnessed the unregulated land use in Metro Manila and the neighboring provinces which caused deforestation of the surrounding watershed areas such as the Marikina and the Laguna de Bay watershed areas. The loss of vegetation has adversely affected the habitat and population flora and fauna in the area, both aquatic and nonaquatic. Low concentration of dissolved oxygen (DO), a common characteristic of the water bodies in Metro Manila, in combination with the presence of toxic substances may have led to stress response in aquatic ecosystem due to toxicity levels.

2.5 Case Studies

2.5.1 Bagmati River Flood, Nepal

Climate change impact on flood frequency was investigated in Bagmati River Basin of Nepal using bias-corrected global climate model (GCM) precipitation output (Mishra and Herath 2015). Bagmati River Basin is an important river basin of Nepal concerning its significance to flood management, and water supply for domestic and irrigation use (Fig. 2.3). Flood events are regular phenomena in the study area causing large human and infrastructural losses. Almost every year, floods in the lower Bagmati region cause substantial damage to infrastructures, human lives and their properties.

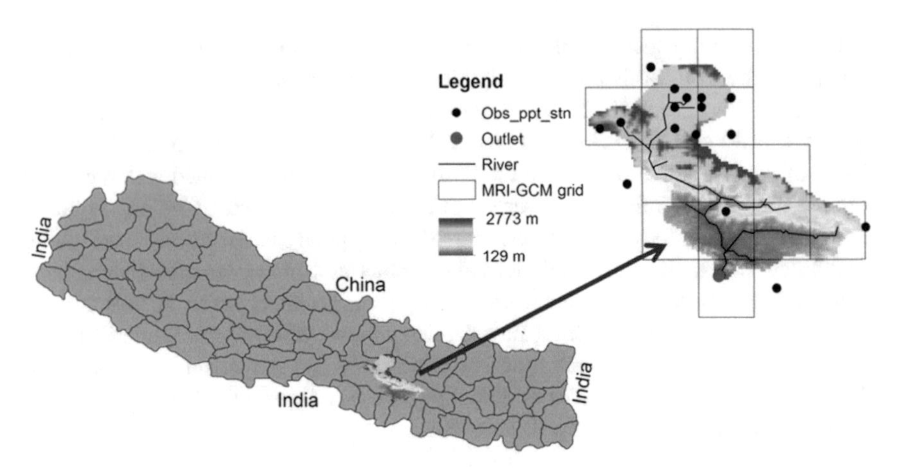

Fig. 2.3 Bagmati River Basin, Nepal

The climate change impact on flood frequency research employed a high-resolution (approximately 20-km) daily GCM precipitation output of Meteorological Research Institute (MRI), Japan. Comparison of observation and GCM data pointed out that the MRI-GCM precipitation output has significant biases in frequency and intensity values. Quantile–quantile mapping method of GCM bias correction was applied for minimizing the biases in precipitation frequencies and intensities. Concept of homogeneous precipitation regions was introduced to link the uneven observation data stations and GCM grid cells. Analyses of return period curves, shape and scale factors at different observation stations enabled delineation of three homogeneous precipitation regions. Accordingly, regional quantile–quantile bias-correction technique was developed for minimizing biases in MRI-GCM precipitation output.

A distributed rainfall runoff model enabled generation of streamflow series using bias-corrected GCM output for 1979–2003 and 2075–2099 periods, as current and future scenarios, respectively. AFFDEF, distributed rainfall-runoff model, enabled generation of daily streamflow series for the available 1979–2003 and 2075–2099 periods, as current and future climates, respectively. Using annual daily maximum streamflow series of the current and future periods, comparative flood frequency analysis was carried out for assessing likely changes in future flood events. Finally, comparative flood frequency analyses were carried out for the simulated annual daily maximum streamflow series of current and future climates. The analyses revealed that the climate change will result in more extreme precipitation events in monsoon months and less precipitation in other months. The analyses also revealed that flood events will be significantly increased in future. The range of change in 2–100 year return period floods was from 24–40% (Fig. 2.4).

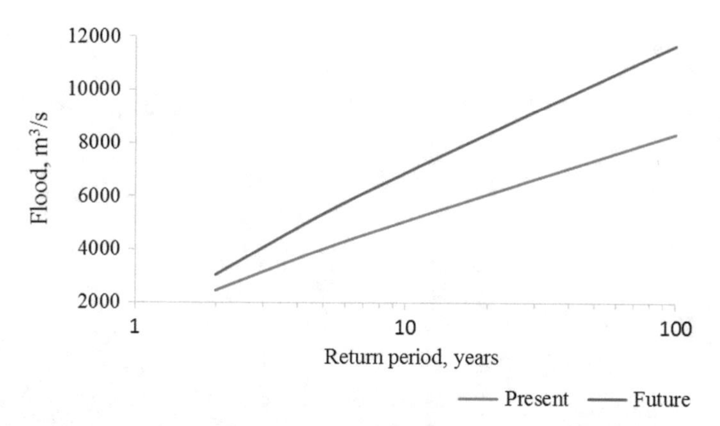

Fig. 2.4 Comparison of annual daily maximum floods for different return periods

2.5.2 Ciliwung River Flood, Indonesia

Ciliwung river basin of Greater Jakarta was investigated as a case study considering frequent flood incidences in the city (Masago et al. 2019; Mishra et al. 2017). Flooding is considered one of the greatest problems that greater Jakarta is currently facing (Fig. 2.5). High flow rates in the Ciliwung River, which flows through the center of Jakarta, regularly cause extensive flooding during the rainy season. This study evaluated flood inundation in the lower Ciliwung River Basin of Greater Jakarta under rapid urbanization and climate change (Fig. 2.6). The future urbanization scenario was based on projected land-use data for 2030. The climate change impact analysis was initiated with comparison of GCM and observation precipitation data, and the results indicated a large bias in the GCM projections. A quantile–quantile bias-correction technique was applied to correct the bias in the high-resolution MRI-GCM projections. Precipitation change assessments over the Ciliwung River Basin were conducted using bias-corrected GCM precipitation data for the current and future climate scenarios. Comparison of 1-day maximum precipitation for 50- and 100-year return period for current and future climate conditions revealed that extreme precipitation events will significantly increase in the future and cause more frequent and larger extreme floods. The HEC-HMS lumped hydrological model was used to simulate the impact of climate and land-use change on the peak discharge in the upper Ciliwung watershed. The peak flow and flood volumes are predicted to increase with rapid urbanization and climate change (Fig. 2.7).

Precipitation output of the MRI-CGCM3, MIROC5 and HadGEM2-ES General Circulation Models (GCMs) with RCP 4.5 and 8.5 emission scenario over periods 1985–2004 and 2020–2039 representing current and future climate conditions, respectively, was used. Similarly, land-use data of 2009 and 2030 were used to represent the current and future conditions, respectively. FLO-2D, a two-dimensional hydrodynamic model, was used to simulate current and future flood inundation simulations (Figs. 2.8 and 2.9). Increasing flood inundation areas and depths (6–31% for

Fig. 2.5 Jakarta skyline after heavy rainfall that caused flood on 10–11 February 2015

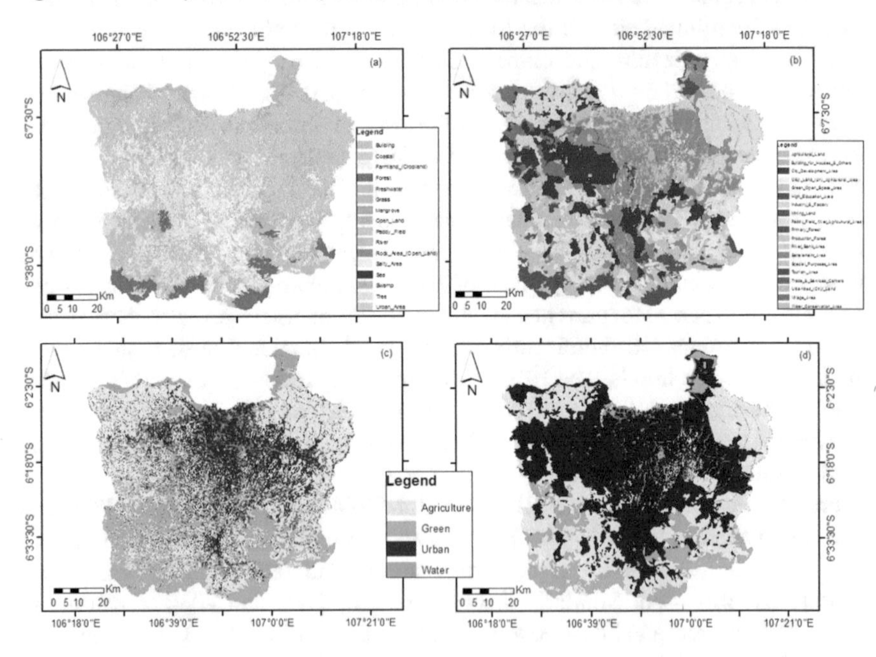

Fig. 2.6 Comparative land-use maps of Greater Jakarta, Indonesia **a** 2009 original **b** 2030 original **c** 2009 derived and **d** 2030 derived

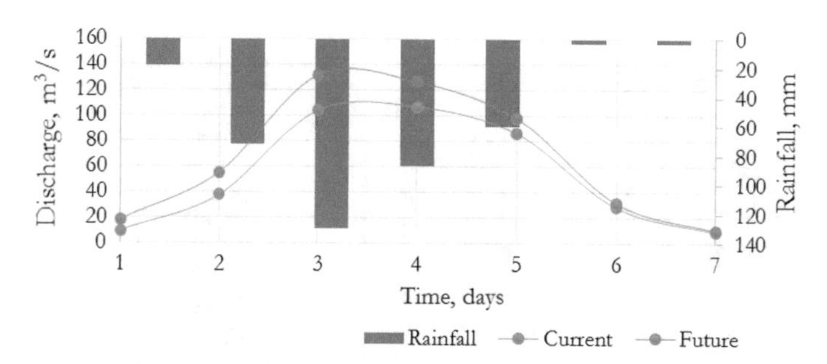

Fig. 2.7 Compartive hydrographs showing impact of urbanization on peak discharge in 2030 from 2007 in Ciliwung River Basin

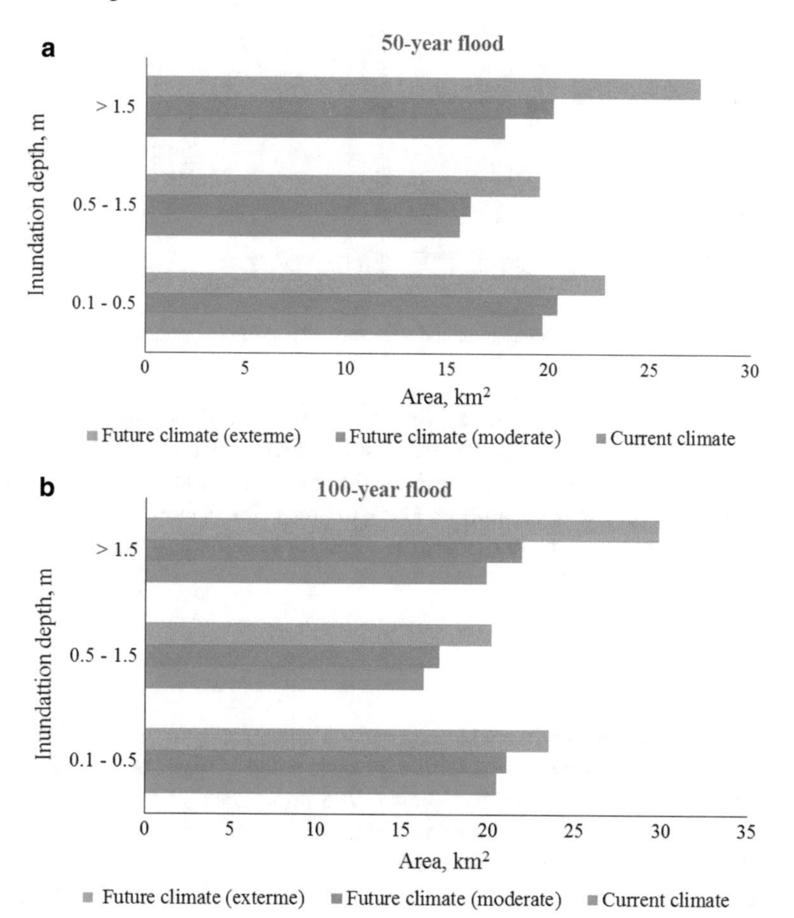

Fig. 2.8 **a** Comparison of flood inundation area for 50-years return period. **b** Comparison of flood inundation area for 100-years return period

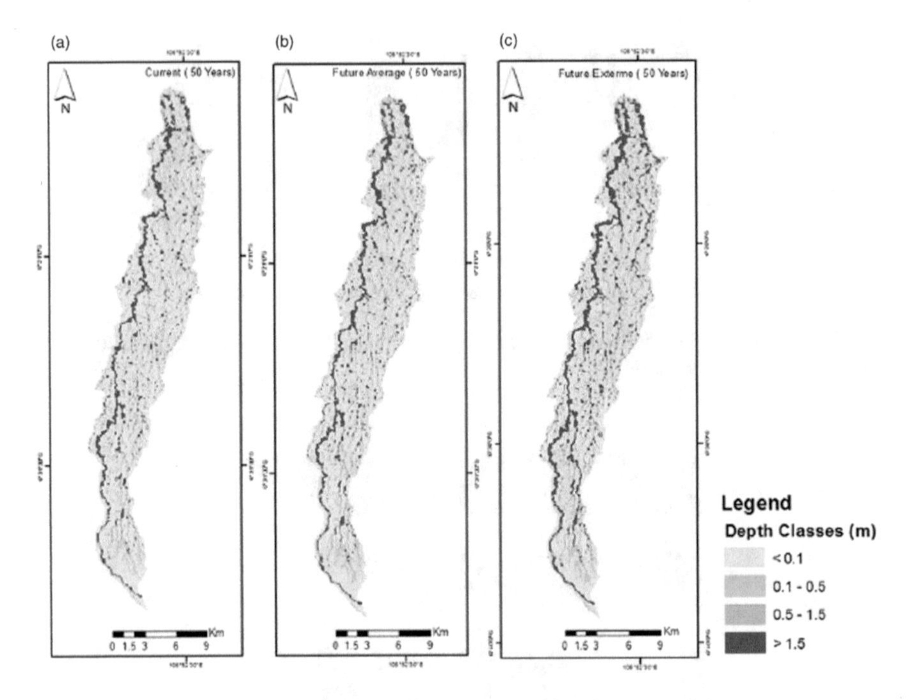

Fig. 2.9 Comparison of flood inundation under the current and future conditions

different GCMs) in the future reveal the need to improve flood management tools for the sustainable development of urban water environments. The flood inundation extent and depths under the future conditions were found significantly higher than those under the current condition. These findings clearly emphasize the need for further flood adaptations and mitigation measures for sustainable urban development.

2.6 Summary

Sustainable development will not be achieved without a water-secure world. A water-secure world integrates a concern for the intrinsic value of water with a concern for its use for human survival and well-being. The importance of this chapter comes from its enhanced understanding of urbanization and climate change on extreme flood events for sustainable urban water management. This chapter was aimed at suggesting policy foresight for infrastructure and land-use management facilitating the need to address challenges to urban sustainability through planning in advance for a population that is projected to come, planning at the local scale anticipating the expected urban growth or resource availability. Therefore, urban expansion through infrastructure development can be demonstrated to contribute to the adverse impacts on ecological sustainability and the high demand of natural resource use within city boundaries. Based on the modeling scenario, protected area zones are considered to constrain change during the modeling process.

The increase in the precipitation and flood pattern will have major implications on the design, operations and maintenance of municipal wastewater management infrastructure. The results of climate and land-use change impacts reflect that the design standards and guidelines currently employed needs revision. Increased peak flows and flood inundation should be considered in future flood management systems, and flexible, adaptive measures should be adopted because of the uncertainty of future climate and land-use changes.

The rapid urbanization coupled with the slow infrastructure development has exerted a tremendous impact on the water resources and urban environment. The increasing population of the metropolis had resulted in the increasing informal settler families. This contributed to the deterioration of the river systems due to untreated waste discharges and increasing incidence of flooding due to the alteration of drainage patterns and waterways. The unplanned expansion of urban areas also led to traffic congestion in the city resulting to increase air pollution and greenhouse gas emission.

Addressing the diversity of issues of urbanization, inclusive economic opportunity, technological innovations, sustainability while still offering a livelihood for the world population has been a great challenge. It has enlightened both the global and local communities to the long-term consequences of our decade-long form of designing cities. That has led to current environmental and inequality related impacts such as water scarcity, air pollution, traffic congestion, cyclones, hurricanes, floods, rampant increase of urban slums, power outages and disease outbreaks. The inequality on who receives better affordable housing and basic environmental services depends somehow on an individual's economic level.

References

Ashley RM, Balmforth DJ, Saul AJ, Blanksby JR (2005) Flooding in the future. Water Sci Technol 52(5):265–274

Bahri A (2012) Integrated Urban Water Management, GWP Technical Background Papers, Global Water Partnership, Stockholm Number 16

Coffman LS, France RL (2002) Low-impact development: an alternative stormwater management technology. In: Handbook of water sensitive planning and design, pp 97–123

Durrans SR (2003) Stormwater conveyance modeling and design. Haestad Press, Waterbury, CT

Grübler A, Fisk D (eds) (2013) Energizing sustainable cities: assessing urban energy. Routledge, Abingdon, United Kingdom

Hatt B, Deletic A, Fletcher T (2004) Integrated stormwater treatment and re-use systems—inventory of Australian practice. Technical Report, Research Centre for Catchment Hydrology, Melbourne

IPCC (2014) Summary for policymakers. In: Climate change 2014: impacts, adaptation, and vulnerability, fifth assessment report of the intergovernmental panel on climate change. Cambridge and New York, NY: Cambridge University Press, pp 1–32

Ismail R (2014) South-East Asian urbanization and the challenge to sustainability: implications for the environment and health. Environ Policy Law 44:55

Jalilov S, Kefi M, Kumar P, Masago Y, Mishra BK (2018) Sustainable urban water management: application for integrated assessment in Southeast Asia. Sustainability 10:122. https://doi.org/10.3390/su10010122

Kefi M, Mishra BK, Kumar P, Masago Y, Fukushi K (2018) Assessment of tangible direct flood damage using a spatial analysis approach under the effects of climate change: case study in an urban watershed in Hanoi, Vietnam. ISPRS Int J Geo-Inf 7:29. https://doi.org/10.3390/ijgi7010029

Kharrazi A, Kumar P, Saraswat C, Avtar R, Mishra BK (2017) Adapting water resources planning to a changing climate: towards a shift from option robustness to process robustness for stakeholder involvement and social learning. J Clim Change 3(2):81–94

Masago Y, Mishra BK, Jalilov S, Kefi M, Kumar P, Dilley M, Fukushi K (2019) Future Outlook of Urban Water Environment in Asian Cities. United Nations University: Tokyo, Japan

Mishra BK, Herath S (2015) Assessment of future floods in Bagmati River Basin of Nepal using bias corrected daily GCM precipitation data. J Hydrologic Eng ASCE, https://doi.org/10.1061/(ASCE)HE.1943-5584.0001090

Mishra BK, Emam AR, Masago Y, Kumar P, Regmi RK, Fukushi K (2017) Assessment of future flood inundations under climate and land use change scenarios in the Ciliwung River Basin, Jakarta. J Flood Risk Manage. https://doi.org/10.1111/jfr3.12311

Mishra BK, Mansoor A, Saraswat C, Gautam A (2019a) Climate change adaptation through optimal stormwater capture measures. APN Sci Bull 9(1). https://doi.org/10.30852/sb.2019.590

Mishra BK, Mebeelo K, Chakraborty S, Kumar P, Gautam A (2019b) Implications of urban expansion on land use and land cover: towards sustainable development of Mega Manila, Philippines. GeoJournal. https://doi.org/10.1007/s10708-019-10105-2

Mitchell VG, McMahon TA, Mein RG (2003) Components of the total water balance of an urban catchment. Environ Manag 32(6):735–746

Oki T, Kanae S (2006) Global hydrological cycles and world water resources. Science 313:1068–1072

Pierpont L (2008) Simulation-optimization framework to support sustainable watershed development by mimicking the pre-development flow regime, North Carolina State University, Raleigh, NC, USA Master's Thesis

Saraswat C, Kumar P, Mishra BK (2016) Assessment of stormwater runoff management practices and governance under climate change and urbanization: an analysis of Bangkok, Hanoi and Tokyo. J Environ Sci Policy 64:101–117

Saraswat C, Mishra BK, Kumar P (2017) Integrated urban water management scenario modeling for sustainable water governance in Kathmandu Valley, Nepal. Sustain Sci. https://doi.org/10.1007/s11625-017-0471-z

Seto KC, Fragkias M, Güneralp B, Reilly MK (2011) A meta-analysis of global urban land expansion. PLoS ONE 6(8):e23777

UN (2015) Transforming our world: The 2030 agenda for sustainable development, A/RES/70/1 (http://www.un.org/ga/search/view_doc.asp?symbol=A/RES/70/1&Lang=E)

UN Agenda 21 (1992) United Nations Conference on Environment & Development. https://sustainabledevelopment.un.org/content/documents/ Agenda21.pdf

UN-DESA (2014) World urbanization prospects: the 2014 revision. Department of Economic and Social Affairs, Population Division, United Nations

UNDP (2006) Beyond scarcity: power, poverty and the global water crisis, Human Development Report 2006, New York

Willems P, Olsson J, Arnbjerg-Nielsen K, Beecham S, Pathirana A, Bülow-Gregersen I, Madsen H, Nguyen VTV (2012) Limitations and pitfalls of climate change impact analysis on urban rainfall extremes. In: 9th international workshop on precipitation in urban areas: urban challenges in rainfall analysis, St. Moritz, 6–9 Dec 2012

Wong T, Brown R (2008) Transitioning to water sensitive cities: ensuring resilience through a new hydro-social contract. In: Proceedings of 11th international conference on urban drainage, Edinburgh, Scotland, UK

Chapter 3
Urban Water Demand Management

3.1 Background

Globally, water demand in urban regions is rising exponentially due to increased economic activities, population growth and rapid rate of urbanization (Immerzeel et al. 2020; Joseph et al. 2020; Fielding et al. 2013). The water providers are confronted with the multiple challenges of limited water availability, increased competition between water users, inadequate water quality due to widespread pollution and rising living standards, resulting in growing water demands in urban regions (Singh and Pandey 2020; Becker 2013). Internal factors such as high operation, energy and maintenance cost of water services, and external factors such as changing climate, changes in frequency and intensity of precipitation introduced uncertainty and complexity in the traditional water supply system (Munasinghe 2019; Poff et al. 2016). The inefficient water supply systems and negative impacts of changing climate (temperature, frequency and intensity of precipitation changes) are threatening water availability and increasing risks of water-related disasters (such as flood and drought) in developing countries (Flörke et al. 2018). Therefore, providing adequate quantity and quality of potable water to rapidly urbanizing regions presents a complex and multi-dimensional challenge for urban water providers (Jensen and Wu 2018).

The previous studies explored the causal relationships between urbanization, population growth and water consumption rate. Many concluded that the rate of water consumption is directly proportional to the population growth and the pace of urbanization (Zubaidi et al. 2020a; Raj 2016; Saraswat et al. 2016, 2017). Notably, in the urban region of developing countries, growing population and increased economic activities exert massive pressure on the locally available water sources (Pandey 2020; Zubaidi et al. 2020a). In this backdrop, increased water demand and exhausted water sources are impacting the efficiency and effectiveness of water infrastructures in an unprecedented manner (Flörke et al. 2018; Saraswat et al. 2017; Wang et al. 2016). In many urban regions, coping with the concerns of growing water demand is perceived from the conventional water supply management lenses

B. K. Mishra et al., *Sustainable Solutions for Urban Water Security*,
Water Science and Technology Library 93,
https://doi.org/10.1007/978-3-030-53110-2_3

(Wang et al. 2016; Herslund and Mguni 2019). In the era of groundwater abstraction, over-exploitation of surface water, inefficient water use and depletion of traditional water sources, the conventional approaches of 'supply over demand' shortfall on the dimension of 'managing demand' (Bharti et al. 2020; Poff et al. 2016). The approach emphasizes augmenting water supply, using engineering practices of building new tunnels to acquire more water supply from other sources (river in other regions), or constructing a new dam (Saraswat et al. 2017; Poff et al. 2016). Undoubtedly, 'supply over demand' is significant in resolving water problems, but it is considered short-term solutions (Singh et al. 2020). Short-term solutions encourage the current practices of the exploitation of local water resources, provide little or no determination to conserve water, hinder awareness regarding water recycling and reuse, and result in a steady increase in water demands in the longer term. The conventional approaches of supply augmentation, overlooking demand management, little incentivization for water conservation and continuin timeworn water policies are leading us toward a scarce water future (Garrone et al. 2020; Li et al. 2020; Hussein 2018).

In this context, to achieve the water security in developing countries, there is an urgent need for transition from conventional water management approaches toward urban water demand management strategies (Herslund and Mguni 2019; Brears 2016). The demand management strategies focus on the 'demand over the supply' (Koutiva and Makropoulos 2019; Saraswat et al. 2017). The approach comprises of socially beneficial policies based on conserving local water sources, encouraging water re-use, recognizing wastewater potential and improving water system efficiency (Pandey 2020). This approach also recommends structural solutions but prioritizes the mitigation or management of water demand over augmenting the water supply (Poff et al. 2016).

3.2 Current Perspective and Prospects

In the current perspective, water scarcity risk is higher in developing countries than in developed countries (Pandey 2020; Rashid et al. 2018). Water scarcity can be defined as a situation when water demands grow beyond the limit of supply potential (Boretti and Rosa 2019; Xinchun et al. 2017). This can be caused due to multiple factors, including limited physical availability of water, unplanned development, infrastructural deficits or governance failures (Boretti and Rosa 2019; Saraswat et al. 2017). Approximately, 16% of the world population is facing water scarcity, i.e., lack of access to affordable and adequate drinking water quality. The developing countries in Asia (mainly south and southeast Asia), Africa and South America are at severe risk of water scarcity (Shah et al. 2018). Recently, different governments and agencies are forecasting the water scarcity doomsday for their respective countries and cities. In 2018, the South African city of Cape Town was in the international news after forecasting the 'day zero' declaration after a series of water shortages and severe droughts (Millington and Scheba 2020; Nhamo and Agyepong 2019). The 'day zero' implies to a situation when the water provider would primarily be unable to supply water, and residents would have to queue for the rationing of water (Millington and Scheba 2020). In 2019, the Indian city of Chennai, Tamil Nadu, declared the 'day zero' situation when the city's water reservoirs will run of water and the water provider would have no water to supply to the city (Ahmadi et al. 2020). An Indian government think-tank estimated and reported that 21 major cities of the country will run out of groundwater by the end of the year 2020 (Desai and Agrawal 2020; Das 2020). Furthermore, the report indicated that the impacts of climate change would exacerbate the water crisis and put the entire country at a water scarcity risk and recommended immediate stronger regulations to curb the crisis (Ahmadi et al. 2020; Vairavamoorthy et al. 2008). In Pakistan, water scarcity and the crisis have led citywide protests in a different city across the country (Ahmadi et al. 2020; Akhtar 2016). In Nepal, the capital city, Kathmandu, is struggling with water scarcity and crisis where residents have to wait in queues for hours to obtain drinking water from the stone waterspouts due to limited capacity of city's water utility to provide/supply water (Pandey 2020; Saraswat et al. 2017). Lack of long-term supply and demand management, absence of forward thinking and holistic planning focuses only on water augmentation projects, and governance inadequacy is responsible for the water crises (Grammer 2019).

The developed countries' urban water management system and water infrastructure are considered adequately established and well structured. The reason behind this is the continuous focus of water providers on long-term supply and demand management, controlled population growth and holistic planning (Tortajada and Biswas 2019; Jensen and Nair 2019; Rathnayaka et al. 2016). The learnings from effective urban water management systems of developed countries can pave the path for urban water demand management strategies. The long-term demand management strategies include reducing water demands by reducing water losses (non-revenue water), water tariff regulations, encouraging water re-use by incentivizing schemes and advancing technologies (Baki et al. 2018). On the other hand, holistic planning aims at coping up with the governance inadequacies by enhancing the coordination and cooperation between water institutions. Singapore is considered a pioneer in water demand management (Jensen and Nair 2019). The water provider in the country can reduce the non-revenue water using the latest technology and infrastructure and encourage conservation by enforcing strong regulations on water tariffs and incentivization (Jensen and Nair 2019; Tortajada and Buurman 2017). In Australia, water providers effectively manage the water demand by promoting rainwater and stormwater harvesting techniques at local and regional levels (Warner et al. 2019; Sapkota et al. 2018). The method of empowering the local communities through government's water awareness campaigns and subsidizing water-saving technologies based equipment to promote recycling of water in urban regions (Warner et al. 2019; Baki et al. 2018). Privatization of the water sector, the approach adopted by the United Kingdom government enhanced the efficiency and tremendously reduced water losses. The private water provider's strategies of using technological advancement such as smart meters, encouraging water reuse and recycling at the household level improved the country's water supply system's efficiency (Bakker 2018). The water providers in the USA analyze the future water demand using integrated scenario modeling and risk analysis in the planning phase. This provides the firms with technical grounding to effectively assess the future demand, operation and maintenance of water services and the water sector's financial sustainability (Zubaidi et al. 2020b). Overall, the urban water management system of developed countries is matured and based on substantial technological advancements. Other measures, such as increasing awareness of water recycling and reuse, catchment and household level rainwater and stormwater harvesting scheme, and managing water demands, are detriminants of urban water demand management (Arfanuzzaman and Rahman 2017).

However, developing countries' water management systems face multi-dimensional challenges of aging and poorly managed infrastructure, relatively new water establishments, excessive leakage and water losses, and ineffective institutions (Raj 2016; Saraswat and Kumar 2016). Furthermore, the lack of capabilities and restrict access to advanced technology present constraints to effective water management (Li et al. 2020; Hussein 2018). We argue here that urban water demand management strategies from developed countries can provide good examples but are not suitable for implementing the same approaches directly to developing countries water management systems. There are various reasons for the argument. First, the developed countries have a robust institutional framework to manage water demand and supply, but in developing countries, water institutions often struggle with fragmentation and silos. The water charges, tariffs policies and levy differ in different regions, provinces, states and jurisdictions (Singh et al. 2020; Sapkota et al. 2018; Raj 2016). Moreover, water institutions have unclear roles or diffused administrative and functional responsibilities. Second, developed countries have robust technological and management systems, but in developing countries, the system is fragile and often based on old technology (Jensen and Nair 2019). Third, in developing countries, water infrastructure is deteriorating and is not managed properly. Rehabilitation and replacements do not often happen, which induces the leakages and losses in the system. The focus of water providers in developing countries is more toward engineering and supply orientation, and this takes bias in planning and designing the system (Sharma and Vairavamoorthy 2009). Lack of human resources and capability for regular maintenances, leak detection, and control, asset management activities, and high non-revenue water directly impacts the efficiency of the water provider (Ahmadi et al. 2020). Finally, the underpricing of water presents barriers to effective water demand management in developing countries due to political ambitions on structural or visible projects rather than long-term measures (Brears 2016).

Reduction in the water demand in developing countries requires a comprehensive approach and holistic planning in urban water management (Vairavamoorthy et al. 2008). Urban water demand management focuses on alternative water availability and sustainable use of water resources. There is a significant gap in the implementation of demand-side management in the local context as the pathway differs in countries' capacity and legal jurisdictions. Based on the differences between developing and developed countries' institutional frameworks and management structure, the procedure for water demand management is presented here. The first step is assessing the current and future water demand based on population growth, urbanization and water consumption rate using projections (Rathnayaka et al. 2016).The second step is to evaluate the current state of water management and distribution infrastructure. Analyzing the water losses (non-revenue water) using internationally recognized indicators is critical in the assessment. After assessing the urban water system, developing new methods for tackling the intermittent water supply is the next important step. Development of appropriate service level and ensuring the reliability of supply will help in the conservation of water. Setting up a system using technologies (e.g., SCADA system), management information system and assets management system are helpful in maintenances and log keeping (Baki et al. 2018; Beal et al. 2016). Depending on the state of water resource in the region, the appropriate structure constructions (such as catchment level rainwater and stormwater harvesting) are also significant (Mishra et al. 2019; Arfanuzzaman and Rahman 2017). From the regulator perspective, reform in water tariffs also provides a way to water conservation and implement fine for wastage and incentive to conserve water (Raj 2016). Finally, increasing awareness of water reuse and recycle, treatment of water at point-of-use and rainwater harvesting using social media, networking Web site, door-to-door campaign, or community reach out are vital activities achieving effective demand-side management. In the next section, we present the strategies and tools for achieving effective urban water demand management in developing countries.

3.3 Urban Water Demand Management Strategies

Achieving urban water demand management has several benefits as it focuses on reducing demand through socially acceptable strategies (Zubaidi et al. 2020a; Beal et al. 2016). It enhances the development and adoption of new technologies, reduces water losses, encourages efficient water use, improves water quality, increases the billing (financial sustainability), maintains the water system's high efficiency and awareness about conservation of water among citizens (Baki et al. 2018). The strategies for urban water demand management focus on technical, non-technical, infrastructural and legal solutions (Stavenhagen et al. 2018; Rathnayaka et al. 2016). The technical solutions include building efficient water systems (treatment, distribution and operational), demand projections using modeling, early detection and leakage control, installing the smartmeter and water-saving devices. Strategies to encourage water recycle and reuse (substitution of water resources with locally available resources), educate community and industry, enhance conservation and creating a management routine of such activities for the water provider are part of non-technical solutions (Stavenhagen et al. 2018; Rathnayaka et al. 2016). Infrastructural solutions comprise the building catchment level rainwater/stormwater harvesting dam, the rehabilitation of old pipes or structure and legal solutions, and the regulations to incentivize water conservation by introducing taxes and water pricing policies. Urban water demand management not only encourages the development and adoption of innovations and technologies but active engagement of community and stakeholders (national and state government, private sector and community) (Stavenhagen et al. 2018; Rathnayaka et al. 2016).

In this research, the urban water demand management strategies classified from two different perspectives, supply-side and demand-side management. Supply-side management strategies are approaches that increase the supply-oriented capabilities of water providers by focusing on technical and infrastructural solutions (Rathnayaka et al. 2016). Demand-side strategies focus on reducing demand using non-technical and legal solutions. Both strategies produce effective results in dealing with the water demand constraints (poor water distribution management, weak water meter management and obsolete infrastructure) (Koutiva and Makropoulos 2019; Saraswat et al. 2017; Rathnayaka et al. 2016).

First, the challenge of reducing the high rate of non-revenue water (30–60% in developing countries) is the most significant issue. To reduce the non-revenue water, distribution network improvement and replacing aging infrastructure (pipes and other) are important. Also, introducing technology (SCADA system) for early detection of leakage and control and regular assessment of the water supply system state and water loss levels using international indicators are used in different countries (Stavenhagen et al. 2018; Norman et al. 2013). Second, achieving a 24×7 water supply to the city reduces the wastewater in storing process and decreases contamination. Third, improving the water supply's reliability by promoting water-saving devices and advancing distribution management ensure water quality (Howe and Smith 2018). Fourth, it is debatable that rainwater harvesting at the catchment level is beneficial or at the household level, but the importance of rainwater harvesting is recognized highly in demand management. It is an alternative to reduce water demand in urban areas and viewed as an emergency (additional) water supply source. Fifth, to reduce the dependability on the surface and groundwater resource, water re-cycle and re-use are vital options, along with the promotion of water-saving devices, installation of smart water meters and efficient system to ensure utilization of drinking at water household and industrial levels (Arfanuzzaman and Rahman 2017). Sixth, the full cost of water pricing, the use of block tariffs, increases awareness regarding water conservation, and other economic incentives such as rebate, subsidies and taxes with strong regulations are essential (Koutiva and Makropoulos 2019). Finally, the education and awareness of water efficiency and conservation are critical strategies in water demand management. Additionally, the institutional arrangements to develop the policy are significant to enhance water demand management (Stavenhagen et al. 2018).

Conventional urban water management based on supply augmentation is not effective in the current context due to the internal and external challenges faced by water providers in urban regions, and water demand management presents an opportunity. The strategies focus not only on new technologies, innovation, reforms and infrastructure development but also on non-technical measures (Baki et al. 2018). It can reduce the burden of water availability and achieve sustainability when the strategies are integrated into planning, design and operation. The local conditions and requirements are fundamental in designing urban water demand management strategies, and its proper implementation supports reduction in water scarcity (Beal et al. 2016). In the next section, the case study from the Kathmandu Valley, Nepal, is presented. The case study evaluates the effectiveness of the supply- and demand-side management strategies in reducing water scarcity in the city. It showed what and how effectively the strategy and measures could help reduce the water crisis in developing countries.

3.4 Case Study: Effectiveness of Urban Water Demand Management Strategies in Kathmandu Valley, Nepal

3.4.1 Introduction

This case study highlights the importance of urban water demand management strategies in achieving water sustainability and security. This case is adopted from 'Integrated Urban Water Management scenario modeling for sustainable water governance in Kathmandu Valley,' first published in the original article category of Sustainability Science journal, Springer in 2017 by the authors (Saraswat et al. 2017). It is relevant to measure the effectiveness of urban water demand management strategies and show its significance in developing countries. The Kathmandu Valley (KV), the capital of Nepal, struggles with water scarcity due to uncontrolled population growth and advancement in economic activities (Saraswat et al. 2017; Shrestha et al. 2015). Urbanization exerting pressure on the local water sources and water supply system results in high water demands and acute water shortages in recent years (Molden 2020; Saraswat et al. 2017; Shrestha et al. 2015). Molden stated that the leaky pipes on the ground and water delivery to the residents through water tankers are common sights (Molden 2020). Moreover, the author argued that the water crisis in the city where nature supplies enough water is not the 'scarcity of water but the scarcity of management.' It is the mismanagement of water resources and inefficient governance of national and local water institutions (Pandey 2020; Saraswat et al. 2017; Shrestha et al. 2015).

Multiple factors were responsible for the water scarcity crisis and unsustainable water use in KV (Mishra et al. 2017, as shown in Fig. 3.1. It is estimated that the city's current water demand is 388.10 MLD (million liters per day) (Saraswat et al. 2017; Shrestha et al. 2015). The water provider Kathmandu Upatyaka Khanepani Limited (KUKL) has limited ability to supply potable water to the city. KUKL is mainly dependent on the available small 30–35 water surface sources (including Bagmati river), 42–45 water reservoirs, 59 operating deep-tube wells (for groundwater extractions) and 39 working pumping stations with 21 functioning treatment plants with an overall capacity of approximately 117 MLD (Chinnasamy and Shrestha 2019; Saraswat et al. 2017; KUKLPID 2015) only. It is analyzed that the water utility capacity to provide potable water to the resident is at 88.8 MLD (33% by groundwater and 67% by surface water) in the dry season and 118.4 MLD (29% by groundwater and 71% by surface) in the wet season (Saraswat et al. 2017; Subedi et al. 2013).

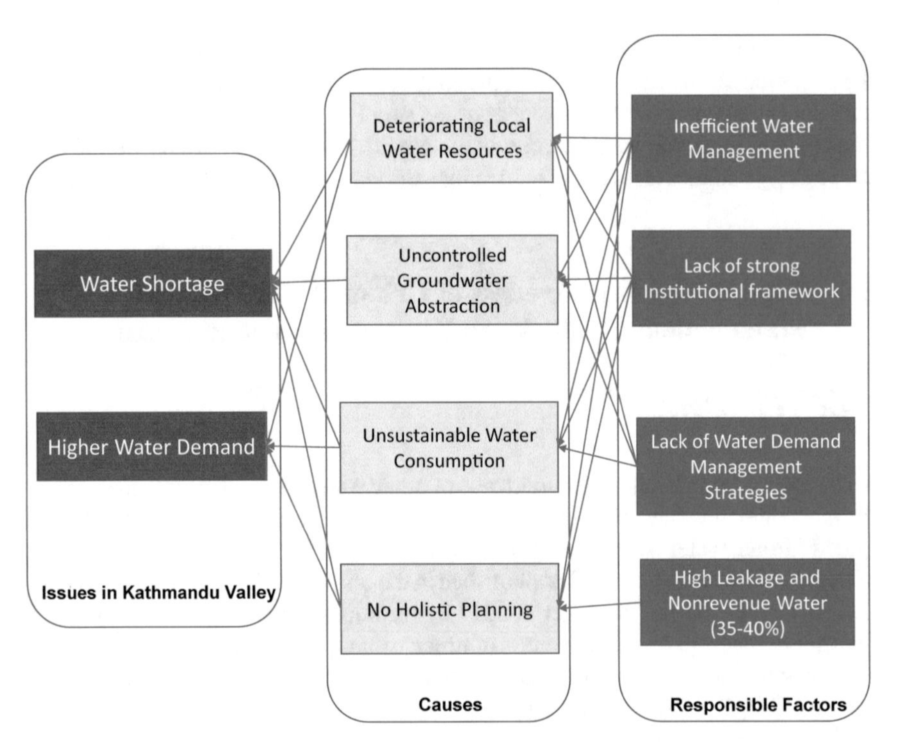

Fig. 3.1 Schematic diagram of issues, causes and responsible factors in KV, Nepal (Adapted from Saraswat et al. 2017)

This shows the disparity in demand and supply of the city. There is a water deficit of 78.5% in a dry season and 63% in a wet season, causing high water demand and water shortages and leading to the water-scarce city. The reasons identified are the non-availability of local water sources to tap for more water (KUKL side of the story). The high rate of non-revenue water (30–40%) due to deteriorating water infrastructure and losses from the existing water distribution systems is a significant factor (Pandey 2020; Shrestha et al. 2015; KUKL 2011). Furthermore, intermittent water supply resulted in uncontrolled groundwater abstraction. The population relied on groundwater, alternative sources of water such as traditional spouts, tube wells and private water tankers to complete the daily needs. Studies showed that the more than 60% of household are dependent on the private wells and use the groundwater despite having piped water connections (Behera and Sethi 2020; Shrestha et al. 2015), which results in groundwater abstraction twice of the rate of natural recharge in the city (Shrestha et al. 2015). Other factors responsible for the situation are identified as the lack of demand-side management approach in the absence of a strong institutional framework and focusing only on augmenting water supply from a nearby river source (Melamchi project).

Recognizing the water scarcity problem of the capital city, the Government of Nepal (GoN) launched 'Melamchi Water Supply Project (MWSP)' an inter-basin water transfer project, augmenting water from the nearby river with the financial help from Asian Development Bank (ADB 2015) in 2000 (Chinnasamy and Shrestha 2019). The project aimed to expand 510 MLD of water supply in three phases (Shrestha et al. 2015). The long delay in completing the project showed the lack of technical capabilities and presence of complexities in the project. The project is still in progress and plans to augment water to the city's water supply system (KUKLPID, 2015) in different phases. In this chapter, this supply-side strategy adopted by water institutions in KV considered a significant approach to achieving water security and demand-side management strategies. The strategies (management options) designed based on the local context to curb the water demand in the city. The scenario modeling approach is used to measure the effectiveness of water demand management strategies.

In this case study, the strategies considered internal and external factors (population, water use and changes in living standards) and focused on demand-side management (long-term approach) along with supply-side management (short-term approach) to combat water scarcity crisis by the year 2030. Finally, the chapter derives multiple scenarios based on management options available in the city to recognize how effective demand management strategies are long-term. The case study identified supply-side management strategies and designed the demand-side management strategies to reduce the water scarcity problem in KV (Fig. 3.2). Along with the comparison of supply-side and demand-side management strategies, the scenario modeling approach analyzes the effectiveness of strategies (Stavenhagen et al. 2018; Saraswat et al. 2017).

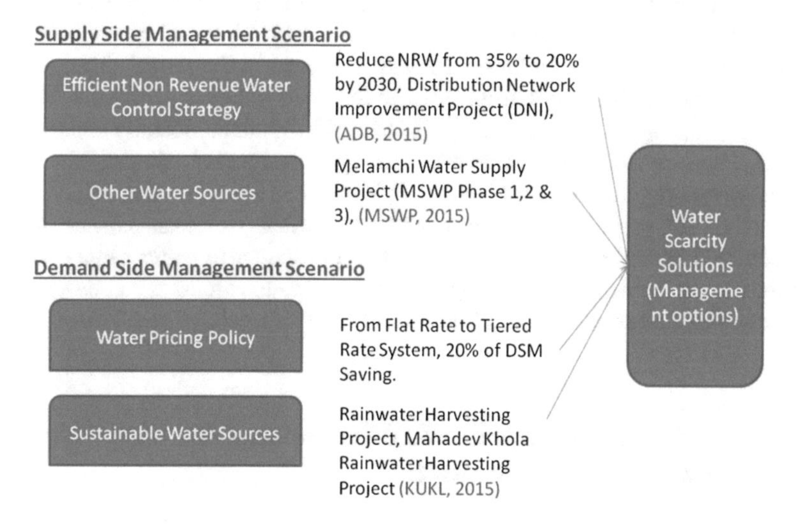

Fig. 3.2 Supply- and demand-side strategies formulation (Adapted from Saraswat et al. 2017)

3.4.2 Methodology

In this case study, urban water demand management strategies are focused on both demand-side and supply-side management solutions. Devised strategies highlight the role of community, stakeholders and institutions to reduce the water demand and scarcity (Hamlat et al. 2013) and use Integrated Urban Water Management principles to design (SEI 2015; Benson and Lorenzoni 2017). To develop the strategies, the literature is drawn from management theories and political science. From the management theories perspective, water provider is considered a firm and reducing wastage to earn profit is an essential part of the strategy. From political science literature, the institutional framework is a significant factor in devising different strategies (Mishra et al. 2017; Saraswat et al. 2017). Data are collected from KUKL annual reports, yearly government reports and master plan for the city, international organization (Asian development bank) report, and seven interviews with KUKL officials. Also, the data on MWSP were collected from KUKL officials and Web site, and based on reports, it is estimated that the three phases of MWSP will deliver water in three stages, 2017, 2022 and 2027, adding 170 MLD in supply system (Bhattarai et al. 2005; MWSP 2015). Finally, along with MWSP, the impacts of population growth, urbanization and climate change on future water demands were assessed to achieve a water-secure future using the principles of IUWM (Saraswat et al. 2017).

To analyze the effectiveness of urban water demand management strategies, water evaluation and planning (WEAP) modeling tool is used. The model incorporates the hydrological parameters and management responses and can analyze effectiveness of various strategies (Yates et al. 2005). The significance of the WEAP model is its analytical capability of designed strategies based on 'What-If' conditions. In the study, multiple strategies are designed and simulated in the model. The model is set up in monthly time steps from 2015 to 2030 for analyzing different (demand and supply) management strategies (SEI 2015). The model efficiently simulates the current state of water supply-demand, water quality situation, allocation priorities and ecosystem requirements (SEI 2015). It is regarded as an efficient modeling tool for practical analysis of urban water management strategy to manage issues such as the reduction in water demands, identification of alternative water supply sources, stormwater use, rainwater harvesting and study of water resources as one system in the literature (Arfanuzzaman and Rahman 2017; Rathnayaka et al. 2016).

3.4.3 Supply- and Demand-Side Management Strategies Formulation and Analysis

Strategy formulation aims to find an approach that can achieve the optimal solution to reduce 100% water scarcity by 2030. With this purpose, multiple strategies are designed, and later stand-alone strategies are integrated into management options (MO1-6). First, the external factors that are responsible for increasing water demand are identified. High population growth (HPG) of 6% per annum and higher living standards (HLS) of using a minimum of 135–150 liters of water consumption (till 2030) were selected as 'High population and high living standard scenario.' In this scenario, the worst-case practices will lead to a water crisis in 2030. The reference scenario is the current scenario, where the current business-as-usual (no radical change in technology and society) practices will lead to a water-scare future. The Figure 3.3 shows that both situations without effective strategies to reduce demand would lead to 'unmet water demand' of 1050 MLD to 700 MLD (approximately). Indeed, this will be disastrous situation for a city in any country.

The supply-side management (SSM) strategies based on 'other sources of water supply (SSM-OWS)' was adopted from the government plan for the MWSP project. The strategy named as MSWP Phase 1/Phase 2/Phase 3-SSM-OWS (other water sources) was based on different phases of adding 170 MLD of water supply. Compared to the reference scenario (current), the results showed that the completion of MSWP phases would positively impact the water scarcity situation in KV. Based on the literature and analysis, it is assumed that the MWSP (all three-phase construction) will be completed by 2027. If the project completed timely, then it is estimated that the 'unmet water demand' in the city will be reduced by 65–50%. This is the representation of the short-term solution approach and decision by the KUKL to solve the water crisis in the shorter term. Another supply-side management strategy is formulated based on the government plan for rehabilitating water infrastructure and distribution networks to reduce non-revenue water (incremental; under 20% by 2030) as it would increase the water supply (SSM-ENRW). The results showed that it would have a positive impact on the longer term. The demand-side management (DSM) strategies are formulated based on the reviewed literature, emphasizing on water pricing policy (DSM-WPP) (tariff reform) to the tired block tariff (estimated that it will save 20% of current expenditure cost) (Stavenhagen et al. 2018; KUKL 2015) and rainwater harvesting estimated saving of 21-37% of water and more. This strategy looks at rainwater/stormwater harvesting as a sustainable water source, titled as DSM-SWS (Arfanuzzaman and Rahman 2017). As shown in Fig. 3.3, all the water demand management strategies effectively reduce water demand and have a positive impact on water conservation compared to the reference (current state of the water supply system of KV) scenario.

Figure 3.3 shows the positive impact of strategies on Kathmandu Valley water supply system but still no stand-alone strategy able to solve water crisis or achieve '100% unmet water demand' by 2030. We took a step further and experimented with same strategies but logically combining different approaches to evaluate the effectiveness of the combination of strategies rather than only one strategy at a time (most likely scenario of real world). To achieve the goal of effectively combined strategies, we tried to evaluate only the supply side, only the demand side, all demand side, but only one strategy from the supply side and vice-versa. This resulted in six different combinations of approaches named as 'management options (MO).' Combination of strategies classified as MO1, ('SSM-OWS (other water sources MWSP))'+'SSM-ENRWS (efficient non-revenue water, i.e., improved water infrastructure to reduce NRW)'+'DSM-SWS (sustainable water sources, i.e., rainwater

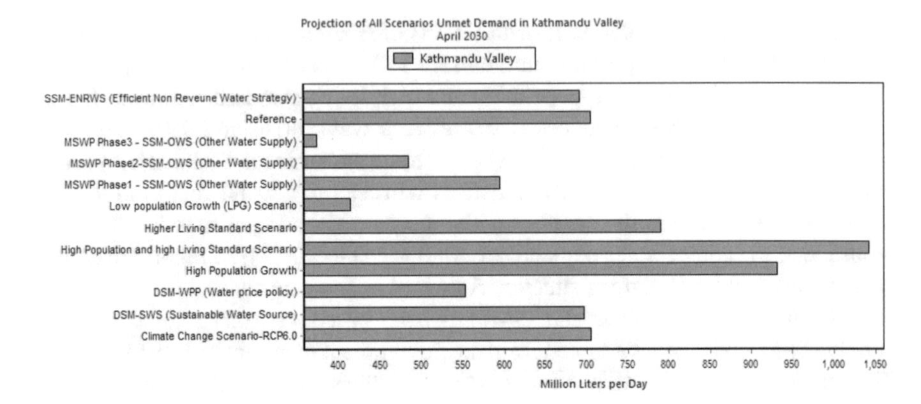

Fig. 3.3 Comparative analysis of individual water demand management strategies for 'Unmet Water Demand in April 2030,' KV (Adapted from Saraswat et al. 2017)

harvesting, stormwater, recycled and reuse of water)'+'DSM-WPP (water price policy change, i.e., change in pricing policy)'), integrating of all designed strategies in one management option. The next is MO2 ('SSM-OWS'+'SSM-ENRWS'), integration of supply-side management strategies together; MO3 ('DSM-SWS'+'DSM-WPP'), a combination of demand-side management strategies only (Saraswat et al. 2017). It is significant in raising awareness and handled at a minimal cost.

The MO4 ('DSM-SWS'+'DSM-WPP'+'SSM-OWS'), combined with supply-side management strategies integrated with the Melamchi Water Supply Project completion phases by 2027; MO5 ('SSM-OWS'+'SSM-ENRWS'+'DSM-SWS'), supply-side management strategies combined with demand-side strategies of introducing sustainable sources of water, i.e., rainwater harvesting, which is essential and perceived as a long-term solution to the problem. Finally, MO6 ('DSM-SWS'+'DSM-WPP'+'SSM-ENRWS'), based on the assumption that MWSP will linger around more due to already many extended deadlines and without the project KUKL, has to cope with soaring water demands. This management option will highlight the case of demand management. The results (Fig. 3.4) showed the comparative analysis of management options 1–6, and the MO1, i.e., all strategies together, is the most effective and optimal strategy to address the water crisis in the city. However, the results showed that all management options (strategies) together will have a significant positive impact on the water system of Kathmandu Valley, instead of being implemented stand-alone strategy (Saraswat et al. 2017).

3.5 Discussion and Summary

Generally, urban water management systems in developing economies are based on conventional water supply-driven approaches. These approaches focus mainly on

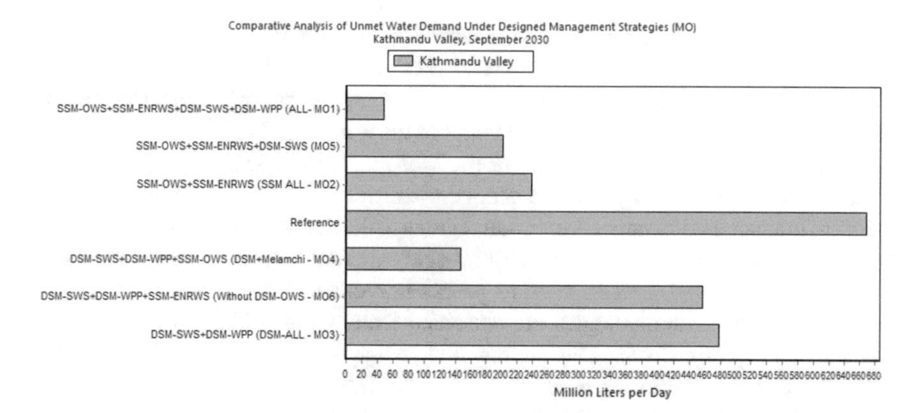

Fig. 3.4 Comparative analysis of management options (combinations of strategies) for 'Unmet Water Demand' in KV, 2015–2030 (Adapted from Saraswat et al. 2017)

augmenting supply in the event of water shortage or a crisis assuming that water resources are infinite. Water providers in the urban region struggle to manage the water supply using such approaches and are confronted with multiple challenges. The limited water availability, inadequate quality of water, rising economic activities and living standards, uncontrolled population growth and changing climate present multi-dimensional and complex challenges for water providers. These challenges (internal and external) exert pressure on the water availability, quality and capability of providers. In the era of groundwater abstraction, over-exploitation of surface water, inefficient water uses and depletion of traditional water sources, the conventional approaches of 'supply over demand' are deemed ineffective. The conventional supply-driven methods of using engineering practices to augment the water supply system are considered short-term solutions. Conventional approaches of supply augmentation using engineering practices are considered short-term solutions while overlooking demand management, little incentivization for water conservation and continuing timeworn water policies leading us toward a water-scarce future. In this background, the shift from supply-oriented management toward demand management becomes vital in devising strategies to achieve water security. Urban water demand management emphasizes on the 'demand over supply' approach and effectively manages the water sector's demand determinants. The water demand management strategies focus on the sustainable use of available (local) water resources, enhancing water efficiency, and recognizing wastewater potential.

This chapter highlighted the potential of urban water demand management strategies in developing countries using a case study. These strategies are based on the latest technologies and innovation to reduce water demand, encouraging reforms to enhance conservation efforts by re-use and re-cycling, and tariff regulations to reduce water use. The infrastructural development encourages rainwater/stormwater harvesting, reduces water losses and incentivizes process (non-technical) development to increase awareness and community participation (Baki et al. 2018; Arfanuzzaman and Rahman 2017). The integration of strategies into planning, design and operation phases of the water system diversifies the water supply and reduces pressure from the water supply management system. Many authors indicated that learning the water management strategies of developed countries could pave the path for water demand management strategies of urban regions in the developing country. The study argued that the local conditions and requirements are fundamental in designing water demand management strategies and its proper implementation.

In this chapter, the strategies are classified from supply-side and demand-side management perspectives. The supply-side management strategies increase the supply-oriented capabilities using technical and infrastructural solutions, and the demand-oriented strategies reduce using non-technical and legal solutions. The combinations of both types of strategies are significant in coping with the constraints (poor water distribution management, weak water meter management and obsolete infrastructure) of urban water demand management in developing countries (Koutiva and Makropoulos 2019). The presented case study evaluated the effectiveness of the urban water management strategies in reducing water scarcity in the context of an urban region in a developing country. It showed what strategies and measures could mitigate water crisis and how effective, they are doing so. In KV, Nepal, multiple water demand management strategies are designed and effectiveness is assessed using the WEAP model. The designed strategies are compared against the business-as-usual practices, and the results showed that unmet water demand would significantly reduce with the implementaion of 'urban water demand management strategies.' Also, the result showed that the strategies would reduce urban water demand, conserve local sources and increase the local water source availability compared to the business-as-usual state of the water management system of KV in 2030. Urban water demand management strategies significantly enhance the transformation of conventional water supply management so as to achieve water security and sustainable development goals (SDG6) in developing countries.

References

ADB (2015) Asian Development Bank Kathmandu Valley Water Supply Improvement Project—Additional Financing: Distribution Network Improvement VII Resettlement Plan. [online] Manilla: Asian Development Bank. https://www.adb.org/projects/documents/kathmandu-valley-water-supply-improvement-project-af-dni-7-rp. Accessed 15 Jul 2015

Ahmadi MS, Sušnik J, Veerbeek W, Zevenbergen C (2020) Towards a global day zero? Assessment of current and future water supply and demand in 12 rapidly developing megacities. Sustain Cities Soc p 102295

Akhtar R (2016) Climate change and geoecology of South and Southeast Asia: an introduction. In: Climate change and human health scenario in South and Southeast Asia. Springer, Cham, pp 1–10

Arfanuzzaman M, Rahman AA (2017) Sustainable water demand management in the face of rapid urbanization and ground water depletion for social–ecological resilience building. Glob Ecol Conserv 10:9–22
Baki S, Rozos E, Makropoulos C (2018) Designing water demand management schemes using a socio-technical modelling approach. Sci Total Environ 622:1590–1602
Bakker K (2018) The business of water. Oxford Handb Water Polit Policy p 407
Beal CD, Gurung TR, Stewart RA (2016) Demand-side management for supply-side efficiency: Modeling tailored strategies for reducing peak residential water demand. Sustain Prod Consum 6:1–11
Becker S (2013) Has the world really survived the population bomb? (Commentary on "How the world survived the population bomb: Lessons from 50 years of extraordinary demographic history"). Demography 50:2173–2181
Behera, B., & Sethi, N. (2020). Analysis of household access to drinking water, sanitation, and waste disposal services in urban areas of Nepal. Utilities Policy, 62, 100996.
Benson D, Lorenzoni I (2017) Climate change adaptation, flood risks and policy coherence in integrated water resources management in England. Reg Environ Change 17(7):1921–1932
Bhattarai M, Pant D, Molden D (2005) Socio-economics and hydrological impacts of Melamchi intersectoral and interbasin water transfer project, Nepal. Water Policy 7(2):163–180
Bharti N, Khandekar N, Sengupta P, Bhadwal S, Kochhar I (2020) Dynamics of urban water supply management of two Himalayan towns in India. Water Pol 22(S1):65–89
Bichai F, Ryan H, Fitzgerald C, Williams K, Abdelmoteleb A, Brotchie R, Komatsu R (2015) Understanding the role of alternative water supply in an urban water security strategy: an analytical framework for decision-making. Urban Water J 12:175–189
Boretti A, Rosa L (2019) Reassessing the projections of the world water development report. NPJ Clean Water 2:15. https://doi.org/10.1038/s41545-019-0039-9
Brears RC (2016) Urban water security. John Wiley & Sons
Chinnasamy P, Shrestha SR (2019) Melamchi water supply project: Potential to replenish Kathmandu's groundwater status for dry season access. Water Pol 21(S1):29–49
Das S (2020) Parched India—A Looming Crisis
Desai M, Agrawal D (2020) The vanishing blue gold—An old problem, a new technology and a big idea—Clensta international. In Socio-Tech Innovation, Palgrave Macmillan, Cham, pp 51–71
Fielding KS, Spinks A, Russell S, McCrea R, Stewart R, Gardner J (2013) An experimental test of voluntary strategies to promote urban water demand management. J Environ Manage 114:343–351
Flörke M, Schneider C, McDonald RI (2018) Water competition between cities and agriculture driven by climate change and urban growth. Nat Sustain 1(1):51–58
Garrone P, Grilli L, Marzano R (2020) Incentives to water conservation under scarcity: Comparing price and reward effects through stated preferences. J Clean Prod 244:118632
Grammer H (2019) Framing the Western Cape water crisis: An analysis of the reporting of five South African publications in 2017 and 2018. Doctoral dissertation, Stellenbosch University, Stellenbosch
Hamlat A, Errih M, Guidoum A (2013) Simulation of water resources management scenarios in western Algeria watersheds using WEAP model. Arab J Geosci 6(7):2225–2236
Herslund L, Mguni P (2019) Examining urban water management practices–Challenges and possibilities for transitions to sustainable urban water management in Sub-Saharan cities. Sustain Cities Society 48:101573
Hussein H (2018) Lifting the veil: Unpacking the discourse of water scarcity in Jordan. Environ Sci Pol 89:385–392
Howe CW, Smith MG (2018) The value of water supply reliability in urban water systems. In: Economics of water resources: institutions, instruments and policies for managing scarcity
Immerzeel WW, Lutz AF, Andrade M, Bahl A, Biemans H, Bolch T, ... Emmer A (2020) Importance and vulnerability of the world's water towers. Nat 577(7790):364–369

Jensen O, Nair S (2019) Integrated urban water management and water security: A comparison of Singapore and Hong Kong. Water 11(4):785
Jensen O, Wu H (2018) Urban water security indicators: Development and pilot. Environ Sci Pol 83:33–45
Joseph N, Ryu D, Malano HM, George B, Sudheer KP (2020) A review of the assessment of sustainable water use at continental-to-global scale. Sus Water Res Manage 6(2):1–20
Koutiva I, Makropoulos C (2019) Exploring the effects of alternative water demand management strategies using an agent-based model. Water 11(11):2216
KUKL (2011) Kathmandu Upatyaka Khanepani Limited. http://www.kathmanduwater.org/home/index.php. Accessed 25 Apr 2015
KUKL (2015) Eight anniversary report. Kathmandu Upatyaka Khanepani Limited (KUKL), Kathmandu
KUKLPID (2015) Kathmandu Upatyaka Khanepani Limited (KUKL) Project Implementation Directorate, Nepal. http://www.kuklpid.org.np/Home/Introduction. Accessed 12 Jul 2016
Li W, Hai X, Han L, Mao J, Tian M (2020) Does urbanization intensify regional water scarcity? Evidence and implications from a megaregion of China. J Clean Prod 244:118592
Millington N, Scheba S (2020) Day zero and the infrastructures of climate change: Water governance, inequality, and infrastructural politics in Cape Town's water crisis. Int J Urban Reg Res
Mishra B, Mansoor A, Saraswat C, Gautam A (2019) Climate change adaptation through optimal stormwater capture measures. APN Sci Bull 9(1). https://doi.org/10.30852/sb.2019.590
Mishra BK, Regmi RK, Masago Y, Fukushi K, Kumar P, Saraswat C (2017) Assessment of Bagmati river pollution in Kathmandu Valley: Scenario-based modeling and analysis for sustainable urban development. Sustain Water Qual Ecol 9:67–77
Molden D (2020) Scarcity of water or scarcity of management?. Int J Water Resour Dev 36(2–3):258–268
Munasinghe M (2019) Water supply and environmental management, Routledge
MWSP (2015) http://www.melamchiwater.gov.np/. Accessed 23 June 2016
Nhamo G, Agyepong AO (2019) Climate change adaptation and local government: Institutional complexities surrounding Cape Town's Day Zero. Jàmbá: J Disaster Risk Stud 11(3):1–9
Norman ES, Dunn G, Bakker K, Allen DM, De Albuquerque RC (2013) Water security assessment: integrating governance and freshwater indicators. Water Resources Management 27(2):535–551
Pandey CL (2020) Managing urban water security: Challenges and prospects in Nepal. Environ Develop Sustain pp 1–17
Poff NL, Brown CM, Grantham TE, Matthews JH, Palmer MA, Spence CM, ... Baeza A (2016) Sustainable water management under future uncertainty with eco-engineering decision scaling. Nat Clim Change 6(1):25–34
Raj K (2016) Urbanization and water supply: An analysis of unreliable water supply in Bangalore city, India. In Nature, Economy and Society, Springer, New Delhi, pp 113–132
Rathnayaka K, Malano H, Arora M (2016) Assessment of sustainability of urban water supply and demand management options: A comprehensive approach. Water 8(12):595
Rashid H, Manzoor MM, Mukhtar S (2018) Urbanization and its effects on water resources: An exploratory analysis. Asian J Water, Environ Pollut 15(1):67–74
Sapkota M, Arora M, Malano H, Moglia M, Sharma A, Pamminger F (2018) Understanding the impact of hybrid water supply systems on wastewater and stormwater flows. Resour Conserv Recycl 130:82–94
Saraswat C, Kumar P (2016) Climate justice in lieu of climate change: a sustainable approach to respond to the climate change injustice and an awakening of the environmental movement. Energy Ecol Environ 1(2):67–74
Saraswat C, Kumar P, Mishra BK (2016) Assessment of stormwater runoff management practices and governance under climate change and urbanization: An analysis of Bangkok, Hanoi and Tokyo. Environ Sci Policy 64:101–117

Saraswat C, Mishra BK, Kumar P (2017) Integrated urban water management scenario modeling for sustainable water governance in Kathmandu Valley, Nepal. Sustain Sci 12(6):1037–1053
SEI (2015) WEAP water evaluation and planning system: user guide for WEAP21. Stockholm Environment Institute, Boston
Shah E, Liebrand J, Vos J, Veldwisch GJ, Boelens R (2018) The UN water and development report 2016 "Water and Jobs": A critical review. Dev Chang 49(2):678–691
Singh V, Pandey A (2020) Urban water resilience in Hindu Kush Himalaya: Issues, challenges and way forward. Water Pol 22(S1):33–45
Singh S, Tanvir Hassan SM, Hassan M, Bharti N (2020) Urbanisation and water insecurity in the Hindu Kush Himalaya: Insights from Bangladesh, India, Nepal and Pakistan. Water Pol 22(S1):9–32
Sharma SK, Vairavamoorthy K (2009) Urban water demand management: prospects and challenges for the developing countries. Water Environ J 23(3):210–218
Shrestha S, Shrestha M, Babel MS (2015) Assessment of climate change impact on water diversion strategies of Melamchi Water Supply Project in Nepal. Theoret Appl Climatol 128(1–2):311–323
Subedi P, Subedi K, Thapa B (2013) Application of a hybrid cellular automaton—Markov (CA-Markov) model in land-use change prediction: a case study of Saddle Creek Drainage Basin, Florida. Appl Ecol Environ Sci 1(6):126–132
Stavenhagen M, Buurman J, Tortajada C (2018) Saving water in cities: Assessing policies for residential water demand management in four cities in Europe. Cities 79:187–195
Tortajada C, Biswas AK (2019) Objective case studies of successful urban water management
Tortajada C, Buurman J (2017) Water policy in Singapore
Tortajada C, González-Gómez F, Biswas AK, Buurman J (2019) Water demand management strategies for water-scarce cities: The case of Spain. Sustain Cities Society 45:649–656
Vairavamoorthy K, Gorantiwar SD, Pathirana A (2008) Managing urban water supplies in developing countries–Climate change and water scarcity scenarios. Physics Chem Earth Parts A/B/C 33(5):330–339
Wang XJ, Zhang JY, Shahid S, Guan EH, Wu YX, Gao J, He RM (2016) Adaptation to climate change impacts on water demand. Mitigation and Adaptation Strategies for Global Change, 21(1):81–99
Warner D, Lewis K, Tzilivakis J (2019) Stormwater harvesting and flood mitigation: A UK perspective. Urban Stormwater Flood Manag. Springer, Cham, pp 29–47
Xinchun C, Mengyang W, Xiangping G, Yalian Z, Yan G, Nan W, Weiguang W (2017) Assessing water scarcity in agricultural production system based on the generalized water resources and water footprint framework. Sci Total Environ 609:587–597
Yates D, Sieber J, Purkey D, Huber-Lee A (2005) WEAP21—a demand-, priority-, and preference-driven water planning model: Part 1: Model characteristics. Water Int 30(4):487–500
Zubaidi SL, Ortega-Martorell S, Al-Bugharbee H, Olier I, Hashim KS, Gharghan SK, ... Al-Khaddar R (2020a) Urban water demand prediction for a city that suffers from climate change and population growth: Gauteng province case study. Water 12(7):1885
Zubaidi SL, Ortega-Martorell S, Kot P, Alkhaddar RM, Abdellatif M, Gharghan SK, ... Hashim K (2020b) A method for predicting long-term municipal water demands under climate change. Water Resour Manag 34(3):1265–1279

Chapter 4
Water Quality Restoration and Reclamation

4.1 Background

On the planet earth, in total, 96.5% of water is saltwater stored in the form of oceans and only 2.5% of freshwater is available for the human survival. The groundwater makes up about 30.1% of available freshwater, which is around 0.61% of the entire world's water (Fig. 4.1) (Gleick 1993, 2000).

Interestingly, the total groundwater availability is almost equal to the total freshwater stored in the form of snow and ice. This shows the importance of groundwater and relevance as natural storage that can buffer against shortages of surface water (Tanvir 2008). United States Geological Survey (USGS) states that there are about 4.2 million cubic kilometers of water within 0.8 km of the planet's surface (Hinrichsen and Tacio 2002). Every year due to the hydrological cycle, more than 500 thousand cubic kilometers of renewable freshwater shifts from oceans to the land in the form of rain or snow, but only 10% of which can be considered accessible for human use (Gleick 2000).

Access of good quality freshwater resources is heavily skewed by rapidly increasing local demand for various purposes, urbanization, aquifer yield and climate change. There is a strong linkage between freshwater and population, which is explained in Fig. 4.2, and the increase in population is directly proportional to the water demand (Vörösmarty et al. 2000). Increased water demand, environmental needs, land-use changes, rapid urbanization, groundwater mining, deterioration of water quality, pollution from local and diffuse sources and impacts on public health are the factors responsible for the severe water quality crisis and of water scarcity in the cities (Saraswat et al. 2016).

Apart from population, climate change scenarios project an exacerbation of the spatial and temporal variations of water cycle dynamics, such that discrepancies between water supply and demand are becoming increasingly aggravated. The frequency and severity of floods and droughts will likely change in many river

B. K. Mishra et al., *Sustainable Solutions for Urban Water Security*,
Water Science and Technology Library 93,
https://doi.org/10.1007/978-3-030-53110-2_4

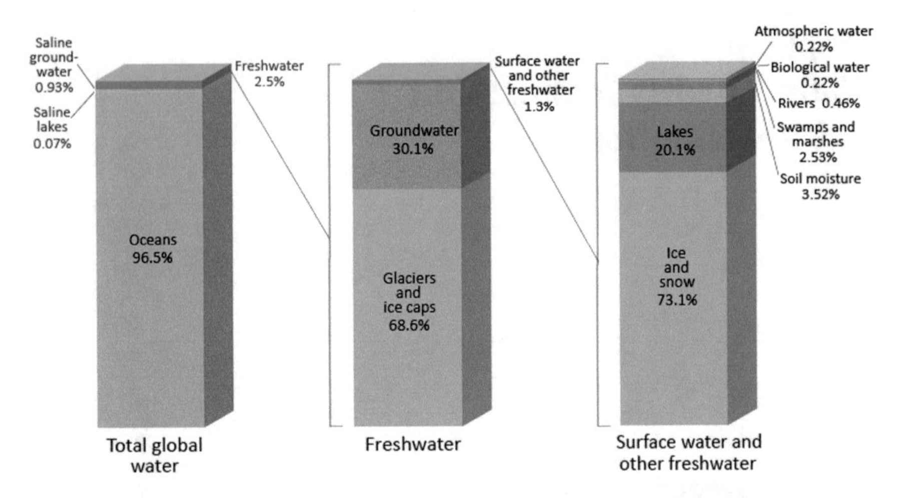

Fig. 4.1 Global water distribution. (*Source* Shiklomanov 1993)

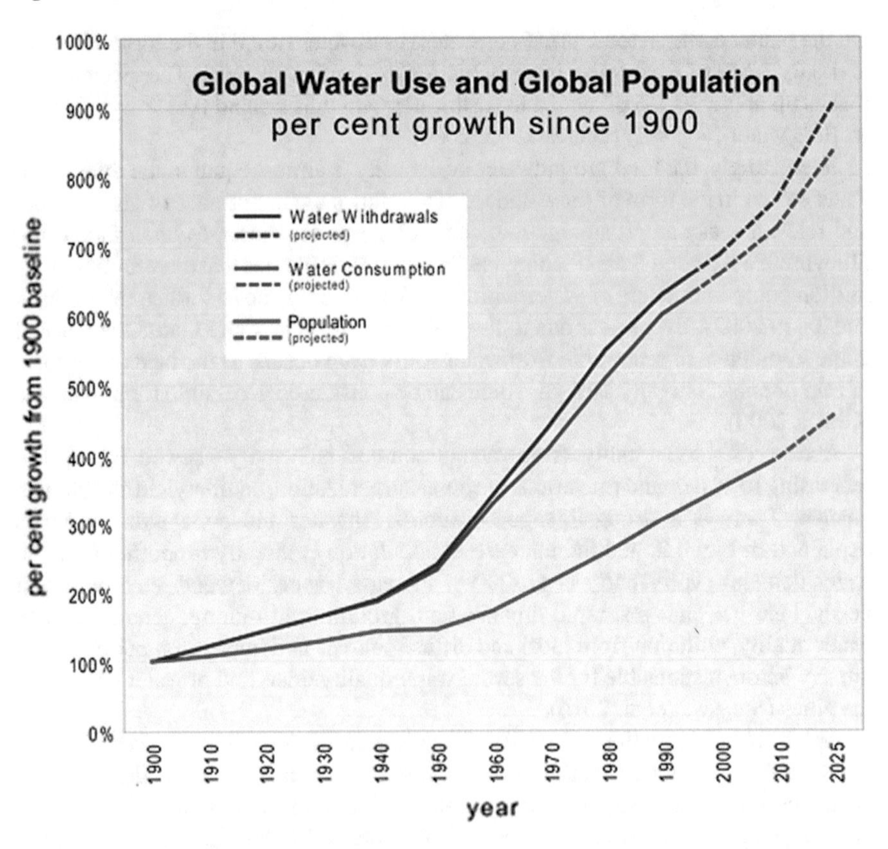

Fig. 4.2 Rate of growth in freshwater withdrawal and consumption has been even more rapid than global population growth. (*Source* Shiklomanov 1999), US Census Bureau (2011)

basins worldwide. Droughts can have very significant socioeconomic and environmental consequences. Two-thirds of the world's population currently live in areas that experience water scarcity for at least one month a year (UN 2015). About 500 million people live in areas where water consumption exceeds the locally renewable water resources by a factor of two. Highly vulnerable areas, where non-renewable resources (i.e., fossil groundwater) continue to decrease, have become highly dependent on transfers from areas with abundant water and are actively seeking affordable alternative sources. The availability of water resources is also intrinsically linked to water quality, as the pollution of water sources may prohibit different types of uses. Increased discharges of untreated sewage, combined with agricultural runoff and inadequately treated wastewater from industry, have resulted in the degradation of water quality around the world. If the current trends persist, water quality will continue to degrade over the coming decades, particularly in resource-poor countries in dry areas, further endangering human health and ecosystems, contributing to water scarcity and constraining sustainable economic development. Reports by the Intergovernmental Panel on Climate Change (IPCC Fifth Assessment Report; Jimenez et al. 2014) have placed emphasis on freshwater resources, particularly their vulnerability because of climate change. The development of management and adaptation measures is therefore critical and should recognize that water resources are fundamental to basic human needs and for facilitating present and future development projects.

As topographical low points in the landscape, aquatic systems collect and disperse water, sediment and heat, integrating changes occurring throughout watersheds. Urban development near streams alters inputs (including nutrients, contaminants and sediment), flashiness of discharge and temperature (Paul and Meyer 2001; Morgan and Cushman 2005). Climate change is also expected to have far-reaching impacts on streams, from altering temperature and runoff regimes to increasing the frequency and intensity of droughts and floods (Milly et al. 2005; Alcamo et al. 2007). It is difficult to predict the combined effects of climate change and urbanization from empirical data because changes are occurring on broad spatial and temporal scales, and typically, conditions cannot be replicated. Yet, anticipating the combined environmental impacts of such anthropogenic changes is critical to developing proactive strategies to protect ecosystems and the services they provide (Kumar et al. 2017, Clark et al. 2002; Walsh et al 2005; Palmer et al. 2008, 2009). Thus, process-based models are instructive tools for investigating these complex stressors in a timely manner.

4.2 Effects of Land-Use Change

Land-use change is a complex process affected by both natural processes and human activities. According to Lambin and Geist (2006), the major types of land-use changes include desertification, deforestation, agricultural land encroachment and urbanization. Urbanization is one of the extreme forms of human-induced land-use change.

It results from the intricate actions of various physical and socioeconomic factors. Urbanization transforms natural landscape to urban impervious land, altering the watershed hydrology and introducing pollutants to natural water bodies. As discussed by Reginato and Piechota (2004), the increase in impervious land from urbanization will lead to an increase in surface runoff, which can carry a larger amount of non-point source pollutants from urban areas to the downstream receiving water bodies. Surface runoff from rooftops, city streets and parking lots has been some of the key factors for non-point source of water pollution in many urban areas (Tong and Chen 2002). Conversely, population growth also increases the demand of freshwater supply. The combined effects of urbanization and population burst will challenge future freshwater availability, especially in arid or semi-arid areas. He and Hogue (2012) noted that under such an environment, urbanization is often the most influential factor in sustainable water development. Research has shown that population growth drives the process of urbanization. For example, Ningal et al. (2008) and Jenerette and Wu (2001) attested that population growth is a crucial socioeconomic driver in urban sprawls. A United States Geological Survey (USGS) study on urban growth of American cities (Acevedo 1999) corroborated that the urbanization process is generally attributed to the increase in population. Therefore, in order to generate better future urbanization and land-use change scenarios, one needs to study how future urbanization is interacting with future population growth in the area. Tobler (1970) developed a demographic model to simulate urban growth. It is the first application of a computer simulation model to explain the relationship between population growth and urbanization. Since then, several studies have applied computer simulation methods to study how urbanization and land-use changes are driven by population growth. Based on these studies, it seems that population-urbanization models based on Geographical Information Systems (GIS) can be effective to simulate future land-use patterns for use as inputs to water quality models.

Generally, both surface water and groundwater provide potable water, as long as it is treated sufficiently (Miller 2006; Bauder et al. 2011). Groundwater is preferred over surface water because of its reliability during droughts or dry season, while surface water sources can deplete quickly. The quality of any body of surface or groundwater is a function of either both natural influences and human influences. Declining water quality has become a global issue of concern as human populations grow, industrial and agricultural activities expand and climate change threatens to cause major alterations to the hydrological cycle.

Water quality protection, water recycling and safe reuse and safely managed sanitation are crucially important in an era when recovering water from the waste streams for use in agriculture and other sectors; as well as, recovering nutrients and energy are seen as multiple revenue streams that enhance the economic value of wastewater. Also, as groundwater is less polluted, it is easier and cheaper to treat it in comparison with surface water. Therefore, in recent past all around the world, people are sustaining mostly on groundwater resources.

Without human influences water quality would be determined by the weathering of bedrock minerals, by the atmospheric processes of evapotranspiration and the deposition of dust and salt by wind, by the natural leaching of organic matter and nutrients from soil, by hydrological factors that lead to runoff and by biological processes within the aquatic environment that can alter the physical and chemical composition of water. Typically, water quality is determined by comparing the physical and chemical characteristics of a water sample with water quality guidelines or standards. Drinking water quality guidelines and standards are designed to enable the provision of clean and safe water for human consumption, thereby protecting human health. These are usually based on scientifically assessed acceptable levels of toxicity to either humans or aquatic organisms.

Globally, the most prevalent water quality problem is eutrophication, a result of high-nutrient loads (mainly phosphorus and nitrogen), which substantially impairs beneficial uses of water. Major nutrient sources include agricultural runoff, domestic sewage (also a source of microbial pollution), industrial effluents and atmospheric inputs from fossil fuel burning and bush fires. Lakes and reservoirs are particularly susceptible to the negative impacts of eutrophication because of their complex dynamics, relatively longer water residence times and their role as an integrating sink for pollutants from their drainage basins.

4.3 Water Quality and Health Nexus

Municipal wastewater is a major source of waterborne human pathogens. Discharging wastewater into the environment without adequate treatment increases the risk of infectious diseases caused by these pathogens. Three targets under Sustainable Development Goals 3 and 6 (Targets 3.3, 3.9, and 6.3) clearly state the need to mitigate waterborne infectious diseases by significantly decreasing the proportion of untreated wastewater (United Nations 2015). Urban flooding and heavy rainfall are often associated with waterborne infectious diseases. Flooding causes municipal wastewater to overflow from urban sewerage, septic tanks and latrines, all of which contain pathogenic microorganisms. Non-proper management of sewerage system and non-availability of sufficient capacity wastewater treatment plants to handle the wastewater being generated locally will exacerbate the criticality of water scarce situation. As it is expected that the climate change would increase the frequency and intensity of flooding in Southeast Asian countries, infectious diseases spread via floodwater would be of great concern especially in urbanized areas.

Lack of proper sanitation and limited access to safe water causes 1.6 million deaths per year worldwide, and from this, most of cases are of developing countries, and the main cause of disease related to drinking water in developing countries is pathogenic viruses, bacteria, protozoa and insects developing on contaminated water. One of the reports predicted that contaminated water is responsible for 15–30% of gastrointestinal diseases. Other diseases caused by consumption of unsafe water are diarrhea, cholera, stomach infection, skin problem, nausea, typhoid fever and legionellosis. Most increasing waterborne disease is typhoid fever that is caused by Salmonella typhi and Salmonella paratyphi, respectively. Even acute viral diseases like hepatitis A and E, rotaviruses and the other protozoa-related diseases caused by Giardia lamblia are often found related to inadequate supply of safe water and sanitation practices (Schwarzenbach et al. 2010). Pathogens like Cryptosporidium parvum, Campylobacter jejuni, enterotoxigenic and enteropathogenic *Escherichia coli*, Shigella spp., or Vibrio cholera cause most of the chronic health diseases (Albert et al. 1999). Pathogenic microorganisms such as *E. coli* and cryptosporidiosis generally cause outbreak of diseases in developed countries, due to microbial distribution in warm water supplies and air condition from houses, complexes and hospitals.

Poor sanitation and hygiene is the leading cause of diarrhea, the second largest cause of death in children under age 5 in developing countries (UNICEF 2012). In addition, many of the negative outcomes that follow from unsustainable sanitation and wastewater management overwhelmingly impact the poor, marginalized and vulnerable and undermine efforts to reduce poverty and discrimination. Improved sanitation and wastewater management systems that prevent exposure of human populations to pathogens and toxic substances can make vast improvements in public health.

4.4 Wastewater: Global Trends

Globally, two million tons of sewage, industrial and agricultural waste is discharged into the world's waterways, and at least 1.8 million children under five years old die every year from water-related disease, or one every 20 s. Twenty-one of world's thirty-three megacities (>10 million people) area are on the coast, where fragile coastal ecosystem area is at risk because of untreated wastewater being dumped there. Even some parts as if Southeast Asia is rich in water resources, being home to 27% of the world's freshwater resources (FAO 2003). However, low-income and lower middle-income countries of Southeast Asia, as classified by UN DESA (2015), face a particular lack of adequate wastewater treatment systems and have poorly built and maintained septic tanks that result in untreated and disease-inducing wastewater being released into open urban water bodies. The region is currently affected by clean water scarcity, as an estimated 80% of all wastewater is discharged untreated directly into rivers, lakes or oceans (UN Water 2017).

On average, high-income countries treat about 70% of the municipal and industrial wastewater they generate. That ratio drops to 38% in upper middle-income countries and 28% in lower middle-income countries. In low-income countries, only 8% undergoes treatment of any kind. These estimates support the often-cited approximation that, globally, over 80% of all wastewater being discharged without treatment. In high-income countries, the motivation for advanced wastewater treatment is either to maintain environmental quality or to provide an alternative water source when coping with water scarcity. However, the release of untreated wastewater remains common practice, especially in developing countries, due to lacking infrastructure, technical and institutional capacity and financing. Due to the differences in the current levels of wastewater treatment overall, the efforts required to achieve SDG Target

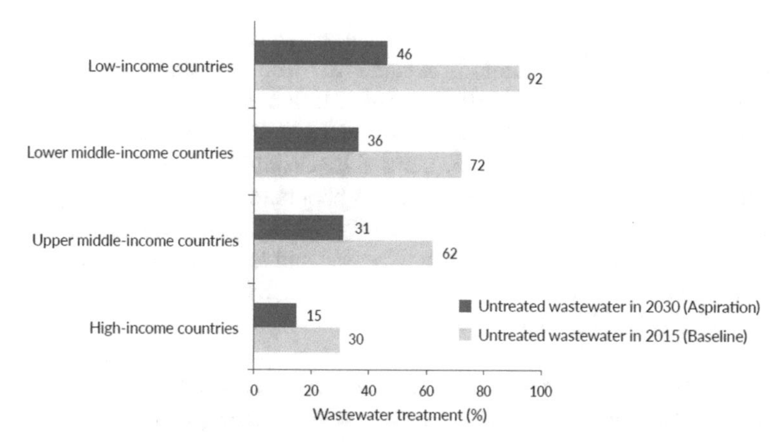

Fig. 4.3 Percentage of untreated wastewater in 2015 in countries with different income levels and aspirations for 2030 (50% reduction over 2015 baseline) (*Source* Sato et al. 2013)

6.3 (related to wastewater management) will place a higher financial burden on low-income and lower middle-income countries (Fig. 4.3), putting them at an economic disadvantage compared to high-income and upper middle-income countries (Sato et al. 2013). More than 50% of human health problems in the world are mainly linked with contaminated water and more people die because of polluted water than are killed by all forms of violence including wars.

The possible main reasons for deterioration of water environment through wastewater generation in the developing countries can be explained by aforementioned points:

(1) Still developing countries rely on the concept of 'Develop first treat second.' It means that the development planner/developers are not operating under same umbrella, i.e., not following properly the policy of urban growth if any given by central government. As a result, they keep extending urban areas without giving proper consideration to connect it through central/sewerage treatment system for their own benefits.
(2) Still significant gaps within the legal frameworks (e.g., regulations and standards) for addressing water pollution issues. Even, in some cases, these water quality standards differ for lakes and rivers and favor certain industries. In addition, measures on water quality are constrained by a lack of organizational, technical and scientific capacity, as well as limited financial resources.
(3) Still in many countries, local residents have to pay some amount in order to connect to integrated sewerage water system also termed as the first initial payment. For example, in Philippines, it is 30,000 PhP/household compared to average income of family which is 20,000 PhP/month (Philippines Statistics Authority 2015).

(4) In terms of technical problem, lack of infrastructure (WWTPs and solid waste management infrastructures) is capable of 100% population coverage (4%—Indonesia, 22%—Philippines), because of lack of funds.
(5) Segregation of gray and black water (unlike in many developed countries)—construction on the top of septic tanks at household levels—Therefore, it is too hard for local government to encourage people to let them clean these septic tanks at the cost of dismantling the facility without giving any other alternative.
(6) Non-awareness about sectoral use of reused and reclaimed wastewater.
(7) Consumer perception—Still in many developing countries, using the food crops produced using the wastewater.

Social acceptance (farmers, retailers and consumers): This is the most sensitive area of this topic. Farmers are not going to reuse water, if their product cannot be sold. Consumers will not buy products where reuse water was used unless it is proven safe. Campaign for the same should be done from some local influential personality and political figures.

Social issues play a significant role in water reuse initiatives and should be adequately addressed. With adequate political willingness accompanied by awareness programmes, these cultural, religious and social objections can be overcome.

Although wastewater is a critical component of the water management cycle, water after it has been used is all too often seen as a burden to be disposed of or a nuisance to be ignored. The results of this neglect are now obvious. The immediate impacts, including the degradation of aquatic ecosystems and waterborne illness from contaminated freshwater supplies, have far-reaching implications on the well-being of communities and peoples' livelihoods. Continued failure to address wastewater as a major social and environmental problem would compromise other efforts toward achieving the 2030 Agenda for Sustainable Development.

In the face of ever-growing demand, wastewater is gaining momentum as a reliable alternative source of water, shifting the paradigm of wastewater management from 'treatment and disposal' to 'reuse, recycle and resource recovery.' In this sense, wastewater is no longer seen as a problem in need of a solution; rather, it is part of the solution to challenges that societies are facing today. Wastewater can also be a cost-efficient and sustainable source of energy, nutrients and other useful by-products. The potential benefits of extracting such resources from wastewater go well beyond human and environmental health, with implications on food and energy security as well as climate change mitigation. In the context of a circular economy, whereby economic development is balanced with the protection of natural resources and environmental sustainability, wastewater represents a widely available and valuable resource.

4.5 Wastewater, Sanitation and the Sustainable Development Agenda

Access to improved sanitation services can contribute significantly to the reduction of health risks. Further health gains may be realized through improved wastewater treatment. While 2.1 billion people gained access to improved sanitation facilities since 1990, 2.4 billion still do not have access to improved sanitation and nearly 1 billion people worldwide still practice open defecation. However, improved sanitation coverage does not necessarily equate with improved wastewater management or public safety. Only 26% of urban and 34% of rural sanitation and wastewater services effectively prevent human contact with excreta along the entire sanitation chain and can therefore be considered safely managed. Building on the experience of the MDGs, the 2030 Agenda for Sustainable Development has a more comprehensive goal for water, going beyond the issues of water supply and sanitation. SDG Target 6.3 states: By 2030, improve water quality by reducing pollution, eliminating dumping and minimizing release of hazardous chemicals and materials, halving the proportion of untreated wastewater and substantially increasing recycling and safe reuse globally. The extremely low level of wastewater treatment reveals an urgent need for technological upgrades and safe water reuse options to support the achievement of Target 6.3, which is critical for achieving the entire Agenda. The efforts required to achieve this Target will place a higher financial burden on low-income and lower middle-income countries, putting them at an economic disadvantage compared to high-income and upper middle-income countries.

The benefits to society of managing human waste are considerable, for public health as well as for the environment. For every US$1 spent on sanitation, the estimated return to society is US$5.5. Overcoming the practical difficulties of implementing water quality regulations can be particularly challenging. In order to realize the goals of water quality improvement and water resources protection, individuals and organizations responsible for various aspects of wastewater management need to comply and act in the collective interest. Benefits are only realized once everyone abides by the rules to protect water resources from pollution. Involving citizens in decision-making at all levels promotes engagement and ownership. This includes decisions as to what types of sanitation facilities are desirable and acceptable, and how they can be securely funded and maintained over the long term. It is especially important to reach out to marginalized groups, ethnic minorities and people living in extreme poverty, in remote rural areas or in informal urban settlements. It is also essential to engage with women, as they bear the brunt of the health consequences stemming from the unsafe management of human waste.

Two major challenges often hinder this process: a lack of reliable, integrated data and a lack of mechanisms that can transform data into useable evidence that is fit-for-policy and that can directly support decision-making.

4.6 Innovations in Water Quality Treatment

Future demographic and economic trends have important implications for household and industry access to urban water and sanitation services. Since demand for global water is projected to increase by 55% by 2050, city dwellers and urban industries are increasingly competing with other water users for access to water resources. If not properly managed, this competition can have undesirable social, environmental and economic consequences. Second, cities are increasingly at risk of floods and droughts, especially as a result of increasing climate variability.

From an institutional perspective, public sector rationalization and territorial reforms (e.g., mergers and amalgamation of administrative regions) have an impact on the allocation of roles and responsibilities in the water sectors and the scale at which water is being managed (local to national). Poor institutional and regulatory framework can have dramatic impacts in some cases sparking vicious circles of under-investment, or favouring expensive technological options at the expense of soft or cheaper options, such as improved water demand managements or environmentally friendly innovations.

Cities can contribute to water resource management, ecosystems and biodiversity conservation, through their design and the infrastructures they rely upon (smart water system, green roofs, more permeable surfaces, etc.). The way in which water is managed in cities has consequences both for city dwellers and for the wider community. Water management in cities dictates water availability (in both quantity and quality) upstream and downstream for other users. It thus also influences the environmental, economic and social development of territories and countries.

The well is but one of a long list of innovations in water technology that have enabled human development to continue apace. Sophisticated pipeline networks and treatment plants today furnish us with this elixir of life and industry. As intense pressure is placed on the planet's limited water supplies, businesses are again turning to technological innovation. New and emerging inventions should see human civilization through the twenty-first century and, with any luck, the next 10,000 years. Nanotechnology in filtration: According to the World Health Organization, 1.6 million people die each year from diarrheal diseases attributable to lack of safe drinking water as well as basic sanitation. Researchers in India have come up with a solution to this perennial problem with a water purification system using nanotechnology.

The technology removes microbes, bacteria and other matter from water using composite nanoparticles, which emit silver ions that destroy contaminants. 'Our work can start saving lives,' says Prof Thalappil Pradeep of the Indian Institute of Technology Madras. 'For just $2.50 a year you can deliver microbially safe water for a family.' It is a sign that low-cost water purification may finally be round the corner —and be commercially saleable.

Membrane chemistry

Membranes, through which water passes to be filtered and purified, are integral to modern water treatment processing. The pores of membranes used in ultrafiltration can be just 10 or 20 nanometers across—3000 times finer than a human hair. But while membrane chemistry has been around for several years, it remains a source of intense research and development. Chemistry significantly contributes to innovative water treatment solutions, such as turning saltwater into freshwater suitable for human consumption.

Recent breakthroughs have been credited with forcing down the cost of desalinated water from $1 per cubic meter to between $0.80 and $0.50 over five years. New ceramic membranes are helping to make treatment more affordable. 'Membrane technology is increasingly important because system integrity, longevity and costs have improved,' explains Paul Street, business development director for engineering firm Black & Veatch.

4.7 Wastewater Reclamation Technologies

Considering the financial, socioeconomic and environmental benefits, wastewater reuse has recorded indisputable progress in recent years in many countries. However, it is highly recommended to have a detail information about the source of wastewater and its usage to make it sustainable. Regarding source or origin of wastewater, manager should know whether this wastewater is gray water (coming from domestic sites) or black water (coming from industrial sites) in order to roughly predict the quality status. On the other hand, it is extremely important to know who is the end user of this wastewater (viz. agriculture, industry or communities) and risks associated with this effluent use. Some of very commonly observed usages are landscape irrigation (e.g., parks, school yards, highway medians, golf courses, graveyard/memorial parks, residential), recreational/environmental usage (e.g., lakes and ponds, marsh enhancement, fisheries), industrial use (e.g., cooling water, boiler feed, process water, heavy construction), not potable urban use (e.g., fire protection, air conditioning, toilet flushing), potable use (e.g., blending in water supply reservoirs as observed in Namibia), groundwater recharge for its replenishment, prevention of saltwater intrusion and subsidence control. However, for groundwater replenishment, one must consider the quality of the recharge water, the physical characteristics of the vadose zone and the aquifer layers, the water residence time and the amount of blending with other sources (Fig. 4.4).

Seawater desalination

Although holding much promise for the future, seawater desalination is still extremely expensive, with reverse osmosis technology consuming a vast amount of energy: around 4 kilowatt hours of energy for every cubic meter of water. One solution being explored in Singapore, which opened its first seawater desalination plant in 2005, is biomimicry—mimicking the biological processes by which mangrove plants and euryhaline fish (fish that can live in fresh briny or saltwater) extract seawater using minimal energy. Another new approach is to use biomimetic membranes enhanced with aquaporin: proteins embedded in cell membranes that selectively shuttle water

Fig. 4.4 An example of wastewater reuse in Salt Lake, Kolkata, India. For Salt Lake in Kolkata, India, since 2005, all the city's gray water (55,0000 m^3/day) is used to fertilize 3000 ha of fishponds, producing some 16,000 tons of fish per year nearly doubling the fish production compared with year 2004 (8500 tons) (*Source* India Water Portal 2013)

in and out of cells while blocking out salts. Harry Seah, chief technology officer for Public Utility Board (PUB), Singapore's National Water Agency, says: 'If science can find a way of effectively mimicking these biological processes, innovative engineering solutions can potentially be derived for seawater desalination. Seawater desalination can then be transformed beyond our wildest imagination.'

Smart monitoring

In developing countries alone, it is estimated that 45 million cubic meters of water are lost every day in distribution networks. Leaks are not only costly for companies, but increase pressure on stretched water resources and raise the likelihood of pollutants infiltrating supplies. 'It does not make commercial sense to invest billions in additional reservoirs and water catchment, treatment plants [and] pumping stations, when as much as 60% of water produced is unaccounted for,' says Dale Hartley, Director of Business Development at SebaKMT, a water leak detection specialist. Different scientific communities also refer this leakage of water as non-revenue water (NRW). This NRW is a big contributor of water scarcity even for the regions blessed with plenty of freshwater resource like Kathmandu, Nepal. Considering the fact that a big chunk of water is getting lost as NRW, professionals working in sustainable water resource management and planning giving special emphasis to reduce it within less than 10%.

The common denominator for all of these social instruments for change is meaningful information. If good decisions are to prevail, then decision-makers need timely access to relevant, understandable information. Without helpful, fit-for-purpose information, then watershed health inevitably suffers chronic degradation. There is growing awareness that the path forward will require less dependence on gray (i.e., concrete) infrastructure solutions and increasing dependence on green (i.e., environmentally sensitive) infrastructure solutions. The common denominator for all of these social instruments for change is meaningful information. If good decisions are to prevail, then decision-makers need timely access to relevant, understandable information. Without helpful, fit-for-purpose information, the watershed health inevitably suffers chronic degradation. There is growing awareness that the path forward will require less dependence on gray (i.e., concrete) infrastructure solutions and increasing dependence on green (i.e., environmentally sensitive) infrastructure solutions.

Several research activities on assessment and adaptation to global change impacts on freshwater resources have been identified:

(1) The development of monitoring networks and databases for change analysis
(2) Methods for change detection, attribution and prediction
(3) Prediction of changes in and vulnerability of groundwater, floods, low flows and droughts
(4) Prediction of groundwater quality degradation and restoration
(5) Assessment of snow, ice and glacier mass balances
(6) Assessment of the impact on sediment transport
(7) Integrated water management for adaptation to global change risk
(8) Policy-related interventions for adaptation.

Intelligent irrigation

Approximately, 70% of the world's freshwater is used by the agricultural industry. Applying a more intelligent approach to water management by deploying precision irrigation systems and computer algorithms and modeling is already beginning to bring benefits to farmers in developed countries. However, while this approach embraces new instrumentation and analytical technologies, innovation comes from a change in mindset that emphasizes the importance of measuring and forecasting. In the old days, there was not so much stress on measuring because it was considered that we had plenty of water. It is a bit of a paradigm switch for the water industry, which like others is used to throwing new engineering developments at problems.

Wastewater processing
Engineering still has its place, however. Many people living in urban areas, even in advanced economies, still do not have their sewage adequately treated and wastewater is often discharged, untreated, into rivers and estuaries or used as irrigation water. New technologies are promising to transform wastewater into a resource for energy generation and a source of drinking water. Modular hybrid activated sludge digesters, for instance, are now removing nutrients to be used as fertilizers and are, in turn, driving down the energy required for treatment by up to half. Henceforth, there is an urgent need for wastewater systems that are more compact, so that new plants can be built in urban areas where land is scarce and for upgrading and expanding extant facilities.

Mobile recycling facilities
An unexpected by-product from the explosion of the global hydraulic fracturing industry has been demand for highly mobile water treatment facilities. Investment is being channeled into reverse osmosis units that will allow companies to treat high volumes of water to extract gas and injected into the subsurface. 'There will be knock-on benefits as products [will be developed] with new applications where the price tolerance is much lower,' says Peter Adriaens, professor of environmental engineering and entrepreneurship at the University of Michigan. Adriaens adds: 'As these technologies develop and learn to treat high volumes of water, we will see cheaper, more potable treatment systems and we will start to move away from massive centralized treatment systems.'

Reinventing water treatment instruments
Traditionally, water resource management has strongly relied on wastewater treatment to ensure that water quality is maintained. However, the historic political approaches to managing water quality are under new pressures, such as micropollution. Micropollution is a complex policy problem characterized by a huge number and diversity of chemical substances, as well as various entry paths into the aquatic environment. It challenges traditional water quality management by calling for new technologies in wastewater treatment and behavioral changes in industry, agriculture and civil society. There still remains a great deal of uncertainty concerning the ability of advanced treatment technologies, such as ozonation or activated carbon, to filter micropollutants and their increased energy needs and costs. In light of such challenges, the question arises of if such end-of-pipe solutions are a sustainable way of ensuring water quality in the future? Moreover, what can we learn from the past? To answer such questions, policy analysis provides an overview of potential policy solutions, analyzes the functioning of policy instruments and evaluates their prospects of solving the policy problem at hand. Therefore, it is very high time when a new framework integrates both the problem dimension (i.e., causes and effects of a problem) as well as the sustainability dimension (e.g., long-term, cross-sectoral and multi-level) to assess which policy instruments are best suited to regulate micropollution. We thus conclude that sustainability criteria help to identify an appropriate instrument

mix of end-of-pipe and source-directed measures to reduce aquatic micropollution (Metz and Ingold 2014).

Financing urban water

This module will provide policy guidance on how governments can effectively meet the financial needs to sustainably manage urban water, and maintain, renew and expand urban water infrastructure. Meeting such needs will require major reforms to improve the economic and institutional framework for water utilities and to enhance the enabling environment for attracting sources of finance and reducing investment needs. This work will help governments to better understand and address commercial, political and institutional issues associated with urban water financing and the contribution of economic instruments and innovative financing mechanisms. The project will collect key examples from selected cities to identify the obstacles to the efficient use of available funding and the mobilization of additional sources of finance, including from the private sector. New monitoring technologies help companies to ensure the integrity of their vast water supply networks. Electronic instruments, such as pressure and acoustic sensors, connected wirelessly in real time to centralized and cloud-based monitoring systems will allow companies to detect and pinpoint leaks much quicker.

Governing urban water

This module will provide evidence on the relationship between governance structures for managing water in cities and the performance of water policy outcomes in terms of access, quality, reliability, equity and sustainability. Relying on an extensive survey across 70 + OECD cities, it will cluster cities according to future urban trends and governance features, map who does what within the water chain and beyond (spatial planning, energy and urban development), draw lessons from good practice in managing interdependencies across people, places and policies and provide guidance on how to overcome territorial and institutional fragmentation. Specific attention will be devoted to pioneer cities with forward-looking adaptive governance strategies to cope with future climate, regional and demographic trends. The intended objective is to support policy coherence and effective water management beyond administrative boundaries and sectoral silos.

Regulating urban water

Countries and cities regulate the dimensions of urban water services (the network, quality, service delivery, pricing, etc.) in different ways. One recent trend is the development of dedicated regulatory bodies for drinking water and wastewater. Building on a survey across 30 dedicated water regulatory agencies and recent OECD work on the governance of economic regulators, this module will shed light on the governance arrangements of these dedicated regulatory agencies in order to promote effective and efficient water service delivery and better responsiveness of service providers to urban population needs.

Eco-innovation and urban water

Government policy has a strong role to play in increasing the amount and pace of eco-innovation in urban water delivery that is critical to the improved management of urban water. This module will focus on the economic and institutional aspects of water-efficient spatial design, smart water systems, distributed water systems and green infrastructure. It will address policy blockages to the uptake of innovations and identify economic and regulatory policies that can encourage water innovations in cities. It will build on case studies from OECD cities, which have managed to retrofit existing infrastructures and incorporate innovative approaches.

Managing the urban–rural water interface

The interdependencies between cities and the broader river basin within which cities are located are critical in terms of the linkages with agriculture, the use of green infrastructure in the watershed, institutional arrangements for watersheds and cities, impacts on biodiversity and economic instruments for managing water allocations. The work will build on the current OECD work on food risk management and groundwater depletion and pollution and on an analytical framework to identify and assess urban–rural linkages and partnerships in functional areas. The module will explore the existing bottlenecks as well as positive drivers for more integrated approaches, at the relevant territorial level and beyond administrative boundaries.

4.8 Case Study: Sustainable Water Environment Management in Jakarta

Several research projects are going on which combine different components of urban water environment to give a holistic picture, which acts as necessary ingredient for policy-makers involved in water resource management. One such research project 'Water Urban Initiative' which is transdisciplinary in nature is being carried out at UNU-IAS, Japan. Here, predictive models are developed to evaluate the water environment in urbanized areas based on computer simulations as shown in figure. Using these simulation models and future projections, the future water environment (e.g., risk of flooding in urban areas, water quality and human health risks due to flooding and water pollution) is evaluated under various scenarios. These scenarios include projected changes in population, land use, climatic factors as well as planned or ongoing construction of flood management and water treatment facilities. Study areas are selected on the basis of rate of urbanization, population growth, status of river, geography and climatic condition of the area. Mathematical tools used for water quality and flood inundation prediction in this study are WEAP and FLO-2D, respectively.

The idea behind running these two scenarios is to obtain a deep insight for possible policy intervention and provide a potential solution for water-related problems. Spatiotemporal simulation and prediction of BOD was conducted for 2015 and 2030 using 2000 as a reference or base year (Fig. 4.5). The simulation results

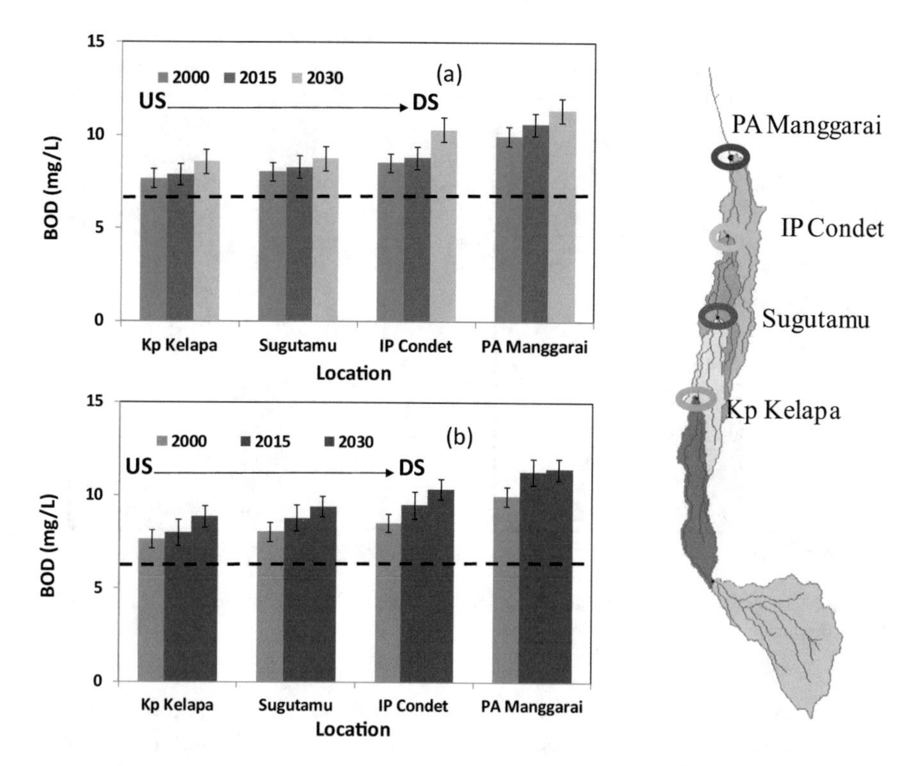

Fig. 4.5 Simulation result of the annual average values of BOD for four different locations in Ciliwung River, Jakarta, for year 2000, 2015 and 2030 **a** scenario considering population growth with planned wastewater infrastructure for 2030; **b** scenario considering population growth and climate change with planned wastewater infrastructure for 2030

for BOD using these two scenarios are shown in Fig. 4.5a, b, respectively. Here, vertical black lines represent standard deviation in data set which is statistically significant. Based on the water quality parameters, the general trend showed that water quality deteriorates from upstream to downstream because of the addition of an anthropogenic output (sewerage). Additionally, the magnitude of deterioration is of higher order in the case of the second scenario where climate change has an additional effect. The reason for this can be the high frequency of extreme weather events due to climate change in the second scenario. Here, extended dry period because of climate change (low flow period in the rivers) might be one of the drivers for increasing the concentration of contaminants. Similar case studies from eight different megacities from Asia can be also found in Kumar (2019). Another powerful approach to help make land-use and climate change adaptation policies more effective at a local scale; Kumar et al. (2020) used a combination of participatory approaches and computer simulation modeling for the water resource management at city level. This methodology also called as the "Participatory Watershed Land-use Management" (PWLM) approach, consisting of four major steps:

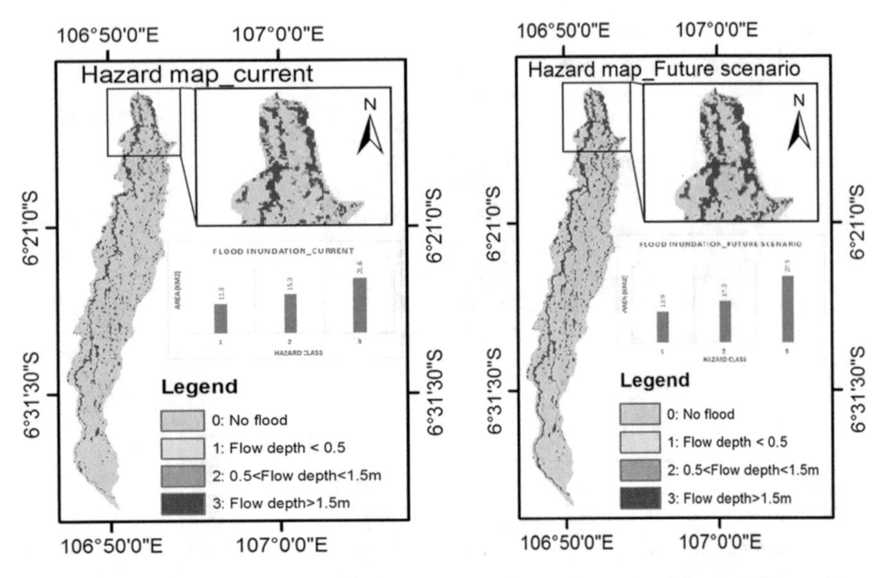

Fig. 4.6 Comparison of flood inundation under the current and future condition in Ciliwung River Basin, Jakarta

(a) Scenario analysis, (b) impact assessment, (c) developing adaptation and mitigation measures and its integration in local government policies and (d) improvement of land-use plan. Detailed analysis and implementation of this methodology are explained for the city of Santa Rosa in Philippines in this paper. Apart of this integrated research, capacity development was also an essential component of this PWLM approach. In order to replicate this methodology and getting the ownership of the research output, capacity development was done in terms of hands on training using freely available numerical tools for the officers involved in water resource management.

Result for the current and future flood inundation is shown in Fig. 4.6. Comparison of 3-days annual maximum precipitation of 50-year return period revealed that there will be a significant increase in extreme precipitation in the future causing frequent and larger extreme floods. Here, it is observed that because of the rapid urbanization and concretization, inundated area especially with flow depth >1.5 m in future time increases. Predicted simulation result of this nature is really useful in term of policy intervention for the local government to adjust their master plan and enhance their water security.

The availability and quality of water are key factors for sustainable industry and households. Moreover, population growth and global climate change are critical issues facing this trans-border region. Changes in temperature have predictable impacts on both water quantity and quality. Economic prosperity and social well-being are both strongly dependent, either directly or indirectly, on water quality and availability. Both small local villages and large urban centers need to be prepared to overcome the risks posed by the scarcity and contamination of water, though some populations are already experiencing shortages in water distribution and drainage services.

Growing challenges resulting from climate change, agricultural intensification, mega-urbanization, resource extraction, energy production and industrialization are resulting in unprecedented pressure on watersheds. Many people despair that 'wicked' problems accumulating to the watershed scale are intractable. Indeed, these problems would worsen indefinitely if our only tools available involved dams, reservoirs, diversions, dredging, channel straightening/hardening, surface drainage and other attempts to manage the flow of water. Fortunately, the root sources of these problems are, primarily, human interactions with water. Managing people to better cooperate within the capacity and resilience of watersheds can be much more efficient, cost-effective and productive than engineering solutions alone to manage the flow of water. The tools available for managing human interactions with water include: awareness, education, public participation, empowerment, social incentive, social license as well as policies, regulations and compliance enforcement.

4.9 Summary

Achieving water security through availability of sufficient water with good quality for all is the main agenda of the United Nation Sustainable Development Goals by the year 2030. The constraints on water availability and deteriorating water quality through global changes threaten secure access to water resources for different uses. The main drivers among these global changes, which are responsible for changes in hydrological cycle, are rapid urbanization, economic development, population burst, land-use/land cover changes and climate changes. Effects of theses global changes can be easily seen through the impacts that water bodies bring to the society in the form of flooding and draught, and most important polluted water bodies result in unhygienic environment/ecosystem and health risks due to emerging pollutants (Mishra et al. 2017). Henceforth, sustainable water resource management is necessary to achieve the goal of water secure future. Despite recent progress in developing new strategies, practices and technologies for water resource management, their dissemination and implementation have been limited. A comprehensive sustainable approach to address water security challenges requires connecting social, economic and environmental systems at multiple scales. This chapter aims to first depict the current situation about water and wastewater especially in Asian countries, followed by brief sketch about outreach and challenges of the existing technologies and policy intervention to achieve sustainable water resource management. Finally, with some successful case studies, this chapter tries to highlight the importance of transdisciplinary research activities which can bring overhaul changes in policy interventions necessary ingredient for achieving water security.

Rapidly increasing global pollution of scarce freshwater resources with largely unknown short- and long-term effects on aquatic ecosystem as well as human health is one of the key problems facing human fraternity. The point and non-point sources of water pollution vary spatiotemporally throughout the world. Key drivers for water pollution include both anthropogenic activities, namely agriculture, mining activities, landfills, industrial and urban wastewater, as well as geogenic releases, namely rock–water interaction. Variability of emerging environmental pollutants and micropollutants posing the greatest health risk and mitigating these water pollution problems is humongous task. Interdisciplinary scientific knowledge and methods designed with local technical, economical and societal dimensions can only solve pollution-related issues at that particular place. Reliable wastewater collection through high sewerage connection rate and treatment systems (centralized or decentralized) is critical for sanitation and for human and ecosystem health. Although, centralized municipal wastewater systems provide reliable solutions to many of these problems but lead to estimated global annual infrastructure costs of US$100 billion over the next 20 years. Such a financial outlay may be problematic issues especially for developing and under-developed nations. Considering the high mortality rate, access to improved sanitation for one-third of the world's population is an urgent issue and should be included as basic human right. Despite the above fact, majority of the financial aid for water-related projects is spent on drinking water supply rather than improving sanitation issues. Recently, because of the race to development and business boost, cheap production in emerging economies is too often accompanied with unacceptable pollution of natural water. International chemical regulation, consumer information and good practice codes should therefore work synergistically to prevent large-scale emission of chemicals into the aquatic environment in all over the globe.

As a future challenge and opportunities, despite the anticipated advances in water treatment technologies, efforts to reduce introduction of emerging chemicals into the aquatic environment should be given highest priority. This requires the improvement of the scientific tools to identify those existing chemicals that need to be substituted and phased out, and the political will enforce such action. In the chemical industry, the 'green chemistry approach' should be more strongly implemented, including efficiency engineering of chemical processes to minimize material flows into the environment and emphasizing the design of new chemicals that are completely biodegradable and therefore of less environmental concern. In addition, improved treatment and removal technologies will allow coping with the legacy of the existing water pollutants. More attention needs to pay to major water polluting sources viz mining activities, landfills, spill sites, agricultural farms, etc., which will continue to threaten our water supplies. Mitigation of these contaminant sources will require enormous financial resources over the next decades and research on effective removal technologies. The high costs of centralized wastewater systems and their low water efficiency require the development of alternative solutions, possibly decentralized systems. They will allow reusing the water and nutrients locally and lead to low discharge systems. The goal of cheap, fast and reliable detection of a broad variety of micropollutants and pathogens in natural water calls for innovative developments in analytical technologies and internationally compatible protocols for water quality assessment. The increasing demand on freshwater resources over the next decades will exert enormous pressure, particularly in arid regions of the world, to protect surface water from pollution. International stewardship for surface water quality will become a high priority to avoid serious water conflicts along international river basins. This chapter indentified sustainable solutions to solve urban water quality issues through wastewater reclamation and restoration, smart agriculture, transdisciplinary research and participatory watershed land-use management approach.

References

Acevedo EH, Silva PC, Silva HR, Solar BR (1999) Wheat production in Mediterranean environments. In: Satorre EH, Slafer GA (eds) Wheat: Ecology and Physiology of Yield Determination, pp 295–331, Food Products Press: Binghamton, NY

Albert MJ, Faruque ASG, Faruque SM, Sack RB, Mahalanabis D (1999) Case-control study of enteropathogens associated with childhood diarrhea in Dhaka, Bangladesh. J Clin Microbiol 37(11):3458–3464

Alcamo J, Flörke M, Märker M (2007) Future long-term changes in global water resources driven by socio-economic and climatic changes. Hydrol Sci J 52(2):247–275. https://doi.org/10.1623/hysj.52.2.247

Bauder TA, Waskom RM, Sutherland PL, Davis JG (2011) Irrigation water quality criteria. Colorado State University Extension Publication, Crop series/irrigation. Fact sheet no. 0.506, p 4

Clark PU, Pisias NG, Stocker TS, Weaver AJ (2002) The role of the thermohaline circulation in abrupt climate change. Nature 415:863–869

Food and Agriculture Organization (FAO) (2003) Agriculture, food and water. Rome, Italy, p. 64

Gleick PH (1993) Water in crisis: a guide to the world's fresh water resources

Gleick PH (2000) A look at twenty-first century water resources development. Water Int 25(1):127–138

Hinrichsen D, Tacio H (2002) The coming freshwater crisis is already here. The linkages between population and water. Washington, DC: Woodrow Wilson International Center for Scholars, pp 1–26

He M, Hogue MH (2012) Integrating hydrologic modeling and land use projections for evaluation of hydrologic response and regional water supply impacts in semi-arid environments. Environ Earth Sci 65(6):1671–1685

India Water Portal (2013) Ingenious system to manage sewage in Kolkata. Accessed 18 Aug 2016 at https://www.indiawaterportal.org/articles/ingenious-system-manage-sewage-kolkata

Jenerette GD, Wu J (2001) Analysis and simulation of land-use change in the central Arizona–Phoenix region, USA. Landscape Eco 16(7):611–626. https://scholar.google.com/scholar?oi=bibs&cluster=9011177689633543304&btnI=1&hl=en

Jimenez BEC et al (2014) Freshwater resources. In: Field CB et al. (ed.), Climate change 2014: impacts, adaptation and vulnerability. Part A: global and sectoral aspects. Contribution of working group II to the fifth assessment report of the intergovernmental panel on climate change (Online), pp 229–269. Cambridge, Cambridge University Press

Kumar P, Masago Y, Mishra BK, Jalilov S, Rafiei Emam A, Kefi M, Fukushi K (2017) Current assessment and future outlook in lieu of climate change and urbanization: A case study of Ciliwung River, Jakarta city, Indonesia. Water, **9**(6):410. https://doi.org/10.3390/w9060410

Kumar P (2019) Numerical quantification of current status quo and future prediction of water quality in eight Asian Mega cities: Challenges and opportunities for sustainable water management. Environmental Monitoring and Assessment, **191**:319. https://doi.org/10.1007/s10661-019-7497-x

Kumar P, Johnson BA, Dasgupta R, Avtar R, Chakraborty S, Masayuki K, Macandog D (2020) Participatory approach for enhancing robust water resource management: case study of Santa

Rosa sub-watershed near Laguna Lake, Philippines. Water, **12**:1172. https://doi.org/10.3390/w12041172

Lambin EF, Geist H (eds) (2006) Land use and land cover change: Local processes and global impacts, Global Change—The IGBP Series, p 222. Springer: Berlin, Heidelberg

Metz F, Ingold K (2014) Sustainable wastewater management: is it possible to regulate micropollution in the future by learning from the past? a policy analysis. Sustainability 6:1992–2012. https://doi.org/10.3390/su6041992

Miller GW (2006) Integrated concepts in water reuse: Managing global water needs. Desal 187:65–75

Milly PCD, Dunne KA, Vecchia AV (2005) Global pattern of trends in streamflow and water availability in a changing climate. Nature 438:347–350

Mishra BK, Rafiei Emam A, Masago Y, Kumar P, Regmi RK, Fukushi K (2017) Assessment of future flood inundations under climate and land use change scenario in Ciliwung river basin, Jakarta. J Flood Risk Manage. https://doi.org/10.1111/jfr3.12311

Morgan RP, Cushman SF (2005) Urbanization effects on stream fish assemblages in Maryland, USA. J North American Benthological Society 24:643–655

Ningal T, Hartemink AE, Bregt AK (2008) Land use change and population growth in the Morobe Province of Papua New Guinea between 1975 and 2000. J Environ Manage 87:117–124

Palmer MA, Reidy CA, Nilsson C, Flörke M, Alcamo J, Lake PS, Bond N (2008) Climate change and the world's river basins: Anticipating management options. Front Ecol Environ **6**:81–89

Palmer MA, Lettenmaier DP, Poff NL, Postel SL, Richter B, Warner R (2009) Climate change and river ecosystem: Protection and adaptation options. Environ Manage **44**:1053–1068

Paul MJ, Meyer JL (2001) Streams in the urban landscape. Annu Rev Ecol Syst 32:333–365

Philippine Statistical Authority (PSA) (2015) Philippines in figures: Databank and information services division. Quezon City, Philippines, p 102

Reginato M, Piechota TC (2004) Nutrient contribution of nonpoint source runoff in the Las Vegas valley. J Am Water Resour Assoc 40(6):1537–1551

Saraswat C, Kumar P, Mishra BK (2016) Assessment of stormwater run-off management practices and governance under climate change and urbanization: an analysis of Bangkok, Hanoi and Tokyo. Environ Sci Policy 64:101–117

Sato T, Qadir M, Yamamoto S, Endo T, Zahoor A (2013) Global, regional, and country level need for data on wastewater generation, treatment, and use. Agric Water Manage 130: 1–13. https://doi.org/10.1016/j.agwat.2013.08.007

Schwarzenbach RP, Egli T, Hofstetter TB, von Gunten U, Wehrli B (2010) Global water pollution and human health. Annu Rev Environ Resour 35:109–136

Shiklomanov I (1993) World fresh water resources. In: Gleick P (ed) Water in Crisis. Oxford University Press, Oxford, UK, pp 13–24

Shiklomanov I (1999) International hydrological programme database, State Hydrological Institute, St. Petersburg, Russia. Accessed 15 Dec 2011 at http://webworld.unesco.org/water/ihp/db/shiklomanov/

Tanvir HSM (2008) Assessment of groundwater evaporation through groundwater model with spatio—temporally variable fluxes: case study of Pisoes catchment. Enschede, ITC, Portugal

Tobler WR (1970) A computer movie simulating urban growth in the Detroit region. Eco Geo 46:234–240

Tong S, Chen W (2002) Modeling the relationship between land use and surface water quality. J Environ Manage 66:377–393

UN (2015) World population prospects: The 2015 revision, Key Findings and Advance Tables: United Nations, Department of Economic and Social Affairs, Population Division

UN Water (2017) Waste water: The untapped resources. Facts and Figure, The United Nations World Water Report. UNESCO publishing: Paris, France, p 83

UNICEF (2012) Pneumonia and diarrhoea: Tackling the deadliest diseases for the world's poorest children. New York, UNICEF, p 77

US Census Bureau (2011) U.S. Census Bureau, International Data Base. Accessed 19 April 2011 at http://www.census.gov/ipc/www/idb/worldpopinfo.php

Vörösmarty CJ, Green P, Salisbury J, Lammers RB (2000) Global water resources: vulnerability from climate change and population growth. Science 289:284–288

Walsh CJ, Fletcher TD, Ladson AR (2005) Stream restoration in urban catchments through re-designing stormwater systems: looking to the catchment to save the stream. J N Am Benthol Soc 24:690–705

Chapter 5
Landscape-Based Approach for Sustainable Water Resources in Urban Areas

5.1 Background

This chapter argues for a landscape-based approach to urban water security, which it asserts as a *subset* of all the ecosystem benefits that can be obtained from ecosystem-based land-use in urban and peri-urban areas. Urban water security is a challenging issue (Fitzhugh and Richter 2004; Nazemi and Madani 2018). Urban areas have caused dramatic changes in the freshwater environments. The most notable alterations are damming, channel modification, stormwater drainage systems, increase in impervious surfaces, highly disconnected waterways (including lakes and ponds), high water demand and increasing pollutants in the water environments (Gurnell et al. 2007; Grimm et al. 2008; Everard and Moggridge 2012). Urban areas also receive food and water from the surrounding peri-urban or rural landscapes, thereby extending their influence into these areas and causing the erosion of multiple ecosystem benefits. By supplying food, energy and material resources to the urban center, the surrounding landscapes also become urbanized, and this situation is especially grave in Asia, in which 14 of the 20 largest cities in the world are located and the region sees high urban agglomerations (Fig. 5.1). A landscape-based approach to water security can simultaneously address multiple problems in this regard. In this chapter, we examine this approach through the interactions between rivers, watersheds and urban environments because rivers are not mere flow of water and also have their own heterogeneous landscapes (Haslam 2008). Unfortunately, this property of heterogeneous landscapes has been disregarded by many landscape ecologists (Wiens 2002), and the role of rivers in urban areas is still inadequately addressed. Hence, there is a need to revisit this concept of rivers having heterogeneous landscape properties in order to conserve them for an ecosystem development of urban areas. Below we lay the foundation of the arguments through visiting some notable examples of urbanization on water resources.

B. K. Mishra et al., *Sustainable Solutions for Urban Water Security*, Water Science and Technology Library 93,
https://doi.org/10.1007/978-3-030-53110-2_5

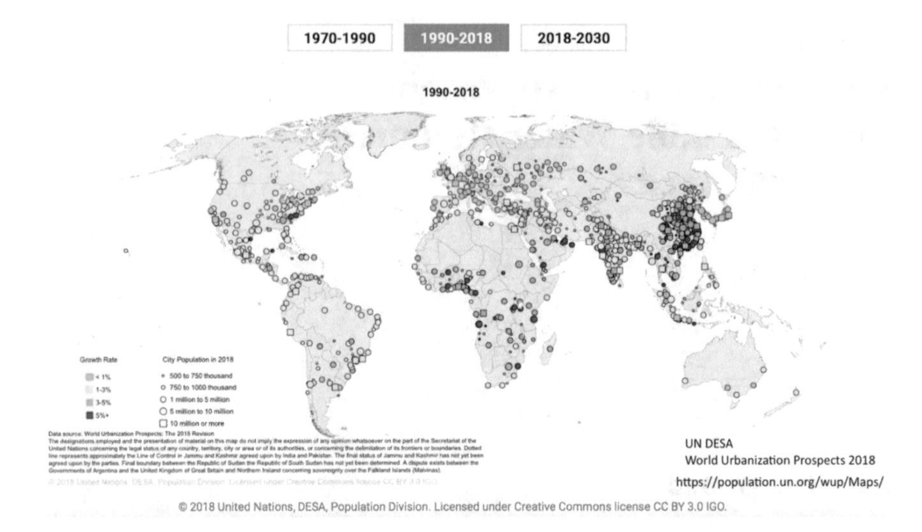

Fig. 5.1 Growth rate of urban agglomerations by size (*Source* UN DESA 2018. https://population.un.org/wup/Maps/)

5.1.1 Effects of Urbanization on Water Resources: Some Notable Examples

Urbanization has multiple impacts on water resources. Structural interventions, such as dams, represent large ruptures in a river system (including the groundwater regime). In the world today, there are about 20,000 dams that have a height of 15 m or more, while hundreds of thousands of other dams are between 5 and 15 m high (Lempérière 2017). Dams prevent rivers to function as environmental systems with multiple ecosystem benefits of the locals (Siciliano et al. 2017) and can have negative consequences such as flooding especially in the period of high rainfall due to dam failures (Lempérière 2017). Furthermore, if we consider the environmental costs of dams, the economic returns they provide may not be that substantial, particularly when the huge costs of maintenance and adverse impacts on the ecosystem benefits are considered (Ansar et al. 2014).

Impermeability is another significant cause of the deterioration of urban rivers. Concretization of bed and banks of a river has adverse effects on the waters that a river carries (Chakraborty and Chakraborty 2017), loss of natural drainage system due to city-oriented developments and increase of impermeable surfaces can decrease ecosystem services in urban areas (Long et al. 2014), and increase urban flooding as well (Esther and Devadas 2016). It has been estimated that urban areas with 50–90% impervious cover can loose 40–83% of the rainfall due to increased runoff (Bonan 2002), thus making urban areas inherently water poor.

Pollution of lotic and lentic systems in and near urban areas is a persistent problem. Dumpsites are inexpensive ways to manage urban waste, especially in developing countries, but they cause considerable damage to urban surface water resources, negatively affecting the health of the urban populace (especially the poor) (Su 2005). In fact, to date, several large dumpsites are massive enough to even trigger landslides, producing man-made disaster risks that threaten hundreds of people (Lavigne et al. 2014; Merry et al. 2005). The problem of groundwater pollution is especially acute in many Asian cities; for example in Delhi, the groundwater is contaminated with sodium, SO_4, nitrate, NO_3 and fluoride, all of which can cause serious health problems, such as cancers and neurological disorders, as well as hampering oxygen transport in the blood (Singh 2010). The Yamuna River that flows through 15 cities in India, including New Delhi, is extremely polluted, and the dissolved oxygen (DO) concentration in certain stretches is below zero due to the dumping of chemical wastes from cities, especially Delhi (Sharma and Chhabra 2015). The resultant effect is, despite being one of the great Himalayan Rivers flowing from glacial headwaters with rich biodiversity, that water from the Yamuna is almost unusable according to environmental standards. The fact that the river is connected to human livelihoods is a cause for serious concern for those who still use its resources. A recent study by Schmidt et al. (2017) suggests that plastic pollution from ten major rivers in Asia accounts for as much as 75% of the pollution loads to the marine ecosystems which also illustrate the state of the rivers in Asia.

Unsustainable freshwater extraction is a major problem in urban areas in highly populated countries (Guneralp and Seto 2008), where large amounts of groundwater are exploited for food production through intensive irrigation. At the same time, these areas pollute groundwater sources, thereby further reducing its usability (Khatri and Tyagi 2014). For example, in the case of Delhi in India, the explosion of the population in multi-storied apartments has increased the demand for water and is stressing groundwater resources; unless checked, the city may become water starved in the near future (Boken 2016). All these suggest that the establishment of new conservation spaces should be considered, such as maintaining the old homes in neighborhoods, as well as homes with gardens that prevent multi-storied apartments from being erected, along with controlling the unabated building that results in sprawl.

Construction has become another major problem. A report by the BBC suggests that rivers in Kenya are being dug down to their bedrock to remove sand for building construction (BBC 2017). These rivers were once lifesavers in dry periods as people dug for water hidden beneath the sand, but now the rivers cannot hold enough water or provide drinking water for villages. More importantly, rivers that lack water-holding capacity and sustained flow cannot provide water for downstream communities, including towns and cities. Sand and gravel mining from riverbeds for construction have been observed to be an unsustainable activity that jeopardizes the physical and biological conditions of rivers (Padmalal et al. 2008).

These factors also pose interlinked problems to water security; for example, dams indirectly create impermeable layers as development in and around the dam site and the city downstream increases due to the need for good road networks as well as housing and industrialization; this development also results in the runoff of pollutants into rivers and waterways.

These issues are among the key examples of the complexities of urban water security; these problems are related to landscape functions that are associated with a river's flow regime (natural flow regime in rivers) and require immediate attention. To rectify these situations, careful planning (with the involvement of citizens and groups affected by urban water problems coupled with environmental education about the linkages between water resource problems and the aquatic environment) is necessary (see Sect. 5.6).

This chapter is divided into seven sections. Following the introduction, we argue for applying the landscape perspective to water security. Section 5.2 discusses landscape issues and the effects of them being decoupled from the society in the urban environments. Section 5.3 addresses the role of ecosystem service (ES)-oriented approaches to sustainable management of urban water resources with specific ES components that can be applied to urban water security. Section 5.4 addresses the issue of landscape diversity and its role in maintaining resilience in aquatic ecosystems; that section is divided into two subsections that integrate areas both within and outside urban domains, primarily because urbanization in the twenty-first century can have a regional effect that is exerted in the surrounding areas (Schneider et al. 2015). Section 5.5 addresses the importance of natural disturbance regimes for maintaining the resilience of the water ecosystem, which is seldom discussed in the resource management sciences. Section 5.6 presents three case studies from Japan, India and the Philippines to illustrate the arguments in the previous sections. Finally, the primary arguments are revisited in the discussion and conclusions (Sect. 5.7).

5.2 Water and Landscape Issues

Reports are not new that extracting freshwater to satisfy the ever-increasing demand for agriculture, industry and urban areas increasingly threatens this rapidly dwindling resource that supports ecosystem health (Postel and Carpenter 1997). Huber (1989) argued that water resource problems continue because most water resource programs focus on human health and not the broad array of natural resource issues. Karr (1991) also identified this reductionist approach as one of the primary causes of failure. It is estimated that by 2030, 60% of the global population is expected to live in urban areas (UN 2012). Freshwater environments depend on the healthy functioning of riverine ecosystems for their long-term use, yet there are few studies that take into account the whole riverine ecosystem management strategies (O'Brien et al. 2016). This implies that restoration and conservation of riverine environments in urban areas (wholly or in part) are necessary for the conservation of freshwater environments. Such riverine environments include components of the freshwater biodiversity, hydrological cycle and their receptors such as different types of vegetation, waterflow through channels and as runoff, groundwater table, lakes, ponds, marshes and soil. Due to the diversity of the aquatic environments involved, there is a need for a landscape-based approach to solve an array of water resource problems.

That the current *approach* to the management of urban water issues is the primary reason that sustainable solutions are not being achieved is not a recent argument.

When landscapes are anthropogenically altered, such as when natural ecosystems that are powered by solar energy are converted to ecosystems that are also powered by petroleum, hydropower or coal-based sources, the resulting ecosystems are not as efficient to provide benefits for a long term because the ecosystems depend on non-renewable energy. Furthermore, these ecosystems have less genetic and cultural diversity, and although they may produce highly complex societies that provide for human well-being (e.g., income, education, health insurance and medical care), they cannot sustain such societies in the long term (consider, e.g., that plants and animals in ecosystems powered by solar energy have persisted for much longer than the existence of humans on earth). Ecosystems that are not only powered by solar energy are mainly located in cities. With twenty-first century seeing greater urbanization than any of its predecessors, urban areas now affect every ecosystem on the planet (McDonald et al. 2009), and in doing so, urban areas in turn become vulnerable to global environmental changes (Lankao and Qin 2011). City environments are ideal places to realize the effects of the $I = PAT$ equation [Impact = population (P), affluence (A) and technology (T)], so urban areas depend on other, surrounding areas for income generation, livelihoods and nutrition. However, this dependence shows the incapability of urban areas to produce and maintain their own provisioning services and in providing supporting and regulating functions of the local (converted) ecosystems (see Sect. 5.4). If urban areas become self-sufficient without further conversion of their landscapes for city uses (or if either the rate of conversion or the possibility of conversion is kept low), much of their ecosystems can continue to provide benefits to the urban society, with water security forming a subset of the 'package' of ecosystem benefits. This is because urban areas have been developed in some of the richest ecosystems on the planet, i.e., fertile river valleys and coastal areas. Many urban ecosystems are supported by a constant supply of freshwater, and more than 50% of the global population lives closer than 3 km to a body of freshwater (Kummu et al. 2011). Unfortunately, despite living so close to freshwater resource, urban areas have disregarded the vital set of ecosystem functions available from the freshwater environments. These functions are either 'locked up' below the soil or in dams and embankments of urban watersheds, or they quickly flow out to sea through concretized and straightened urban water channels without being used by the ecosystems. Additionally, the failure of these functions to enter the 'uncompromised' hydrological cycle deters the release of their benefits to the urban society, which can affect water ecosystems that are not immediately downstream, as well as separate freshwater ecosystems.

An example is the engineered river basin developments in the Kuma River system in Kyushu in Japan, whose downstream channels and ecosystem connectivity have deteriorated. Two dams, Arase and Setoishi, were built in the late 1940s and late 1950s, respectively, to compensate for power shortages in the Kumamoto Prefecture. The electricity generated was transported to the Kitakyushu urban and industrial area in Fukuoka Prefecture. These two dams were part of a set of seven dams built across the Kuma River, primarily to tap its hydropower potential. However, the dams are now

Fig. 5.2 Removal of woody debris from the Setoishi Dam (left). The Yohai barrage, which intercepts the annual upriver migration of sweetfish as well as downstream flow of sediments and organic matters vital for coastal mudflat at Ariake Sea (Photograph by the author, taken on September 5, 2014)

thought by locals to have caused a cascading effect of altering the natural functions of the river that have impacted the downstream communities of Yatsushiro City and its coastal ecosystems. The dams not only stopped the annual migration of sweetfish and other riverine fish species, such as Japanese eels, that used healthy riverine environments to breed and spawn but cut off the vital nutrient and sediment inputs to the Ariake Sea mudflat, which is the largest mudflat in Japan (Fig. 5.2) (Chakraborty and Chakraborty 2017). This has disrupted local fisheries and livelihoods as well as the lifestyles of the downstream communities, affecting different ecosystem services connected to the freshwater environments. Japan opted for nation-wide river basin development plans as part of its post-war urbanization, primarily building dams for flood control, irrigation, hydropower and to supply water for urban and industrial hubs (McCormack 2007; Waley 2000) that saw its river basin undergo swift degradation of biological diversity.

Although Japan is water-secured country, most of its rivers are dammed, often multiple times, and this has seriously disregarded the different services from the freshwater environments. For example, the areas surrounding Tokyo have undergone drastic changes; traditional rural landscapes have declined at the expense of city growth, which has two faces; one, the actual increase in city area with population growth, and two, the rural depopulation that results from people being drawn to the city. The urbanization of Tokyo has 'eroded' traditional lifestyles that depended on the conservation of waterscapes near homes around Tokyo (Takeuchi et al. 2003; Duraiappah et al. 2012). Despite its high degree of development and sophisticated lifestyles, Tokyo is now one of the most water-stressed cities in the world (McDonald et al. 2014).

5.3 The Role of the Ecosystem Service Approach to Water Security

The use of ES-based assessments in natural resource management in urban areas is increasing (Gómez-Baggethun et al. 2013; Milliken 2018). This implies that an ecosystem service (ES)-based approach can be a valid tool for addressing the security of urban water environments. The mainstreaming of the notion that ecosystems offer us various benefits began with the Millennium Ecosystem Assessment (MA) and the Convention on Biological Diversity (MA 2005; CBD 2010), but important theoretical perspectives such as that by Costanza et al. (1997), which attempted to provide the first comprehensive global picture of ecosystem values through an ES framework, appeared well before the MA. However, the MA placed greater effort in grounding such knowledge at the landscape level and related the notion with multiple human well-being components (Fig. 5.3). The MA defines ecosystem services as the '…benefits people obtain from ecosystems' (MA MEA (Millennium Ecosystem Assessment) 2005, p. 40), and it has become a cornerstone on which many other studies have been based. The MA was followed by similar frameworks such as The Economics of Ecosystems and Biodiversity (TEEB), the Intergovernmental Panel for Biodiversity and Ecosystem Services (IPBES), and the Common International Classification for Ecosystem Services (CICES) (TEEB 2010; IPBES, n.d.; Haines-Young and Potschin 2011).

Among these frameworks, TEEB concentrated on the biodiversity components of ecosystems to create sustainable economies by 'valuing' nature through the quantification of limited sets of values that are related to human well-being. TEEB therefore, serves as a market-based, economic approach to ecosystem services. With its detailed typology of provisioning, regulating, supporting and cultural services, CICES added abiotic factors as another important component of ES science, whereas IPBES elevated ES to national and regional policy-making platforms at a much-needed time. First, the MA was aging, and second, there was a lack of efforts by the policy-makers for grounding the concept of social ecological systems (SES) in human society (Saarikoski et al. 2018). IPBES was formed to capture various values of ecosystems that arise from diversity in the human conceptualization of nature based on cultural attributes. Thus, the IPBES aims for the integration of available pluralistic values and an open approach to the perceptions of such values by different societies (Pascual et al. 2017). Several ES are related to cultural identity, spiritual interactions and sense of place, and these need very different approaches in order to be captured (Chan et al. 2012a, b). These studies are constantly fine-tuning the

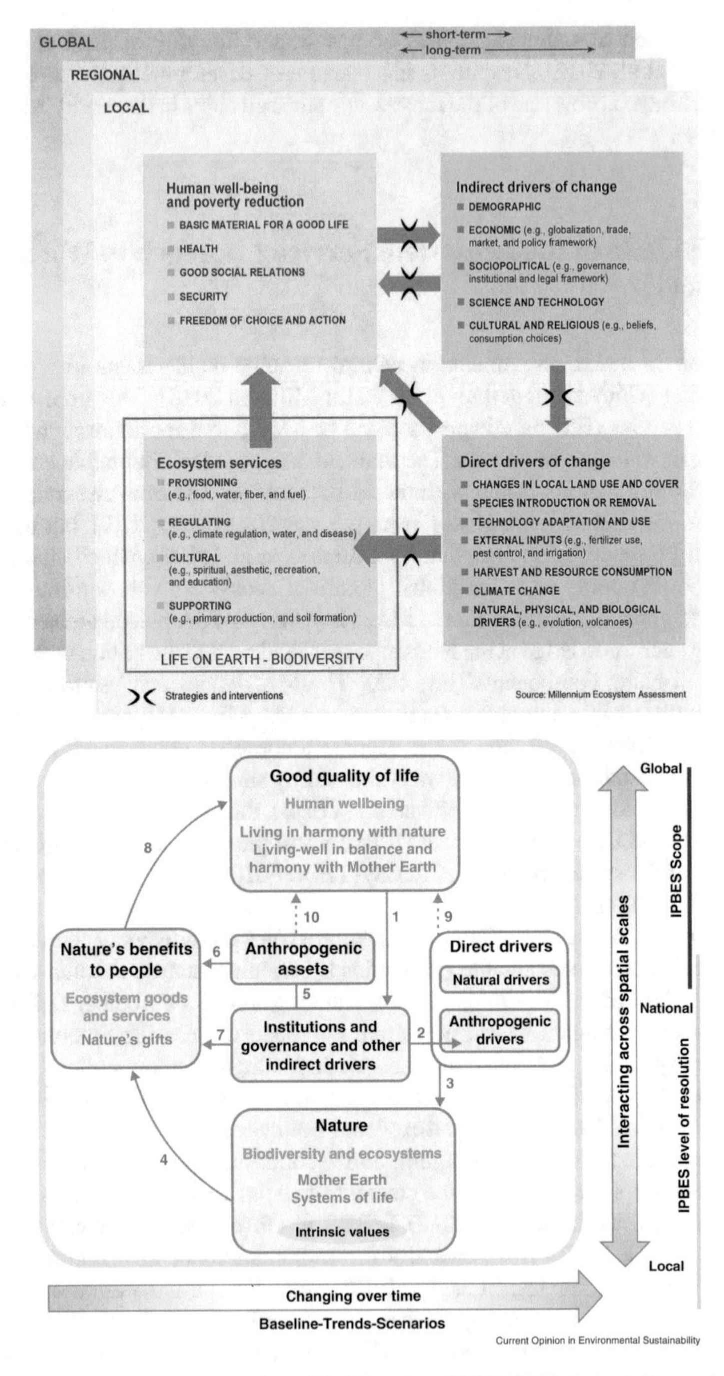

Fig. 5.3 Conceptual framework of MA 2005 (Above) and IPBES 2015 (Below) (*Source* MA 2005, Diaz et al. 2015)

methods of ES application as well as notions of ES-oriented resource management at the landscape level. Although more than 15 years have passed since the MA, ES approaches are still not well applied, urban water management is among the areas that can benefit from the application of this concept (Livesley et al. 2016; Garcia et al. 2016). For global sustainability goals such as SDG 2030, utilizing the evolving concepts of ecosystem service science in order to develop resource-rich, sustainable cities is worth examining.

5.3.1 Ecosystem Service Components that Can Better Address Urban Water Security

The notion of ecosystem services can be helpful to achieve sustainable solutions to an array of different natural resource problems. However, the ES framework is also a contested tool, and certain cautions should be taken when applying it at the landscape level such as a watershed. We may still see greater use of market-based instruments, exploitative uses and commodification of nature even with the rise of ES valuation (Schröter et al. 2014), and especially, this reductionist application of ES can exist in urban areas, where human well-being can be achieved at the expense of synergies between ES. In general, people tend to enjoy *more* ecosystem services where their actions are coupled with the landscape, and in such cases, cultural and intangible interactions with the landscape are vital. Unfortunately, these cultural interactions are less studied and less known, and there are few attempts to bring them into the policy arena. One major problem may lie in the old methods applied in ecosystem service science, such as market-based and quantification techniques to assess 'valuable' ecosystems. However, since this cultural perspective is relatively new, and due to our overwhelmingly linear thinking about the relationships between people and ecosystems, the ES framework can currently only capture a handful of the highly complex and dynamic relationships that exist between people and nature. Thus, while a cultural perspective in ecosystem science is necessary, but it is difficult to incorporate, especially in urban areas that are run by market-based, bureaucratic governance systems. However, if grounded, even partially, by urban planners and developers, it can address some vital ES components of urban areas, such as water security as a component of the whole set of benefits. Urban landscapes can secure water either directly (viz. less evapotranspiration and loss due to runoff over impermeable surfaces) or indirectly (conservation of green water through urban home gardens or watershed protection that supports urban areas). Below, we touch on a few key ES benefits related to a landscape-based management approach in the urban areas.

Urban areas can very easily avail themselves of some particular ecosystem services such as rainwater harvesting, which can be done with limited infrastructure. Rainwater harvesting is also necessary in urban areas where the unmet water demand is high, and rainwater can especially meet drinking water standards as it is regarded as a relatively safe source of water (such as from rooftops) that is away from contamination sources and pathways (Moseley 2005). Water harvesting on a household basis can also help grow home gardens in urban areas, providing a connection to sustainable food production (Panyadee et al. 2016), and agriculture in urban areas can address water security issues, as heavy input-oriented, mechanized and irrigated agriculture remains a major factor for the deterioration of water resources (Elmi et al. 2002).

Urban rivers and riverine landscapes that follow healthy ecosystem functions can enhance in-stream biodiversity, and river restoration through environmental flow (EF) has long been discussed as a meaningful way to maintain ecosystems with their carrying capacity. Environmental flow can benefit watershed ecosystems as it can enhance physical habitat, maintain evolutionary paths that species have followed in adapting themselves to the available natural flow regime, maintain the population viability of species through longitudinal and lateral connectivity and deter the success of invasive species in the riverine ecosystem (Bunn and Arthington 2002).

Urban landscapes that maintain healthy ecosystem functions can gain benefits in the form of climate regulation, pollination, erosion control, temperature regulation and pollution reduction. Practicing agriculture and gardening in the urban soil complex or restoring roadside grasslands can help the population of pollinators near the urban domain, facilitating food production (Bon et al. 2010; Hopwood 2008). Agriculture and aquaculture on impervious surfaces (e.g., rooftops) can also benefit household provisioning and reduce dependency on high input-demanding industrial agriculture (Orsini et al. 2014). However, urban agriculture, if attempted at the industrial scale, can fall into the same economic pitfall of profit maximization and use of polluted environments to produce crops. Therefore, it is smarter to produce at the household level, to avoid negative environmental externalities common in market-oriented agriculture. However, these non-market mechanisms are currently concentrated toward rural or poor urban neighborhoods (Morton et al. 2008). Nevertheless, urban agriculture remains an important sustainability tool to be incorporated into the economic system (Kamiyama et al. 2016). Urban green spaces in addition to more soil moisture retention than impervious surfaces can influence local microclimates in a positive way by reducing the air temperature in summer, reducing fossil fuel-based energy consumption and the resultant urban heat island (UHI) effect (Oberndorfer et al. 2007; Bowler et al. 2010; Jenerette et al. 2011). Urban green spaces can also reduce air and water pollution (Wolch et al. 2014; Escobedo et al. 2011). Forests in urban watersheds can be of immense ecological value (see Sect. 5.4.2), and the so-called green dams (meaning the effects of forests in stabilizing river basins and providing diverse ecosystem benefits such as water preservation, landslide and flood prevention) can be much more beneficial than concrete dams in the long run (Fujiwara 2009). Cultural (recreation, aesthetics, sense of place, social interaction and cognitive development) (Lee and Maheswaran 2011; Haq 2011; Dadvand et al. 2015) and health (Jennings et al. 2016) effects of ecosystems that contribute to better human development should be maintained in urban areas. These can provide positive feedbacks to sustainable water use as poor health and poverty are related to the overexploitation of water resources and pollution (The World Bank 2009). This contribution of better human development in urban areas also ensures a win-win situation for urban water management.

However, one of the major problems for ES-based landscape management is the limited number of studies on praxis. Mainstreaming ES-oriented approaches requires a holistic landscape-based approach rather than strategically dealing with only water-related issues. For example, present river basin and coastal management structures are oriented to averting damage from natural 'disasters' such as floods; but floods are also natural events. Rather than the physical phenomenon of flooding, it is poverty, the unaffordability of health insurance, or floodplain occupancy due to the increasing population in riparian areas that make flooding a disaster (Adhikari et al. 2010). A non-ES-oriented approach may continue to regard flooding as a disaster and attempt to confine the river system via engineering projects primarily, preventing the flow of multiple ES goods and services from freshwater environments. However, an ES-oriented approach may adopt a different perspective, encouraging freshwater flow as a central part of the landscape elements that are fundamental to life (including human life) and increase resilience in urban environments (Table 6.1 in Chap. 6 of this volume). Unfortunately, ES-oriented landscape management is still in its infancy, particularly in urban environments. However, studies do suggest that the investment costs in ES-based approaches for landscape management are less than the benefits obtained from such ecosystems (Karelva et al. 2011).

5.4 Landscape Diversity as a Tool for Urban Water Management

That ecosystem stability depends on diversity is not a new notion (Odum 1953; McArthur 1955); therefore, more diverse ecosystems will have greater stability and be able to withstand more 'shocks,' such as floods, droughts, climate change, pests, and diseases. This diversity is a part of long-term resource security. In attempting to make water-secure urban ecosystems, this diversity must be maintained in the landscapes in and around urban areas (thereby functionally incorporating urban areas into diverse ecosystems and their components).

5.4.1 Areas Inside the Urban Domain

The urban domain refers to the immediate urban surroundings and the jurisdictional areas that are directly involved in city land use, planning and management. Urban parks/urban green spaces are increasingly being recognized as providers of ecosystem services such as water quality (Berkowitz et al. 2003), reduced temperatures, (Bowler et al. 2010), CO_2 assimilation (Niemela et al. 2010), aesthetic benefits, stress reduction and mental health (Chiesura 2004; Conway 2000). All these can also

Table 5.1 Components of an effective urban river restoration

1.	Water quality must be good to maintain wildlife/ecology
2.	Where flooding is already occurring, involve people in the design of the scheme
3.	Involve the community early in planning relating to the river
4.	No further building on the floodplain
5.	Where there must be concrete, attempt to make it look interesting
6.	Optimize opportunities for education about the river and wildlife
7.	Where redevelopment is planned, take the opportunity to remove buildings from the floodplain
8.	As much as possible, allow the river to find its own course, encouraging a meandering shape and allowing the water to flow freely
9.	Where feasible, increase the opportunity for safe access to the river, accounting for the needs of children, the disabled and the infirm
10.	Where land is available and conditions are appropriate, consider creating natural wetlands
11.	Where access is feasible, enhance safe opportunities for people to relax, walk and recreate
12.	Enhance the variety of river flow, depth, shape and ecology
13.	Aim for no litter in the river or along its banks

Source Petts and Gray (2006)

provide economic benefits as well. For example, Luttik (2000) finds that a view of a water body increases house prices in the Netherlands. At the landscape level, planning for urban green spaces that prioritizes river restoration can take place at urban watersheds. Table 5.1 is redrawn from Petts and Gray (2006), and the main point is internalizing the ecosystem (physical/biological) aspects of urban rivers (especially points 1, 4, 6, 7, 8, 10 and 12), while re-orienting human behavior to support the agenda (points 2, 3, 5, 9, 11 and 13).

5.4.2 Areas Outside the Urban Domain

Areas outside the urban domain include peri-urban and rural to forested surroundings—geographical spaces that are dealt specially in this chapter (see Sect. 5.6). These are not directly influenced by urban land use, but they are affected by factors as population increase or a particular agenda for urbanized growth. Such areas provide vital supporting and regulating functions in the catchments of cities that make freshwater available for the urban populace.

The watershed of New York City has been well known for its catchment management, which provides clean drinking water for urban areas. When the Catskill watershed was degraded due to soil and water contamination from fertilizer and pesticide inputs, city developers had to decide between watershed restoration or a water treatment plant that would have required an investment of several billion dollars. The planners decided for the first option, buying land and restoring the watershed, which not only saved several billion dollars but also added other invaluable services such as carbon sequestration and biodiversity conservation (Chichilnisky and Heal 1998). This case suggests that investing in natural capital such as upland forests can be far more beneficial (both economically and environmentally) as it provides bundles of ecosystem benefits that are not available through investment in physical or complementary capital. However, physical or complementary capital (in this case, facilities for buying land) is needed to bring water ecosystems into our current market economy. This matter of interplay of natural and complementary capital has been discussed in Sect. 5.6.

Martin-Lopez et al. (2012) report from a survey of 3379 stakeholders on the Iberian Peninsula that rivers and streams and urban and dryland ecosystems are the least-preferred suppliers of ecosystem services. This shows that these areas are under greater threat of conversion and change, which would further deteriorate their ecosystem services. Additionally, geographic characteristics such as urban areas in the drylands, urban streams or dryland streams face greater threats than urban areas near forested, wetland and coastal areas.

5.5 Natural Disturbance Regimes in Urban Water Security

As noted above, most urban areas occupy prime river aggradation areas such as floodplains. These places are dynamic, as they are shaped by natural disturbance regimes such as floods, and sediment inflows through floods that replenish nutrients to the soil. Unfortunately, rivers are not allowed to flow through their natural course in urban areas, quite understandably because of problems related to space and the 'hazards' posed by floods to life and property. However, despite all these structural adjustments, flooding in urban areas does not show a downward trend (Sala 1992; Rahman et al. 2016).

This is where serious effort is required to understand the trade-offs among the ES benefits available from complementary capital (i.e., man-made infrastructure, machines, buildings, etc.) and natural capital. Pathways for internalizing natural disturbance regimes in river courses should be incorporated into urban landscape planning, and again, this is where the landscape-based approach is crucial. Urban areas have totally altered, straightened and concretized river courses, so their ecosystem benefits have been seriously compromised, only to gain a false sense of security from flooding. On the other hand, the least-altered watercourses will have greater biodiversity and thus more ecosystem benefits as well as more (but infrequent) 'risks' of flooding incidents. These flooding incidents are necessary because natural disturbance regimes that shape and foster ecosystems are needed to maintain complex adaptive systems (CAS), where 'patterns at higher levels emerge from localized interactions and selection processes acting at lower levels' (Levin 1998). A dominant feature of such systems is nonlinearity (Levin 1998). Ecosystems that maintain such a complex adaptive state can be more resilient. For example, regarding sustained water resource availability, if we compare storm hydrographs for rural and urban catchments, we see markedly different water flow regimes. In rural settings, base flow, through flow, groundwater flow and surface flow level off in the hydrograph when they reach the river channel, and this spreads the whole storm event over a longer time, which is not possible for urban watersheds (see Chap. 6). This phenomenon occurs due to the presence of forested areas and agricultural areas, species diversity that maintains such forests and agricultural lands, soil conditions and pervious surfaces, and reduced evapotranspiration. Thus, urban watersheds lack (1) protection against flooding without costly infrastructures and (2) the vital ES benefit of water availability for a longer time. Urbanization in the rivers of Long Island, New York, has been shown to have reduced the base flow by as much as 65% in sewered and 11% in unsewered areas, and thus, urbanization has allowed for a new equilibrium to be established in the urbanized streams (Simmons and Raynolds 1982).

5.6 Case Studies

Critical evaluation of gaps in land-use planning and consideration of successful practices are two approaches that are vital to achieve urban water security; these are exemplified through two case studies below:

5.6.1 Small-Scale and Holistic Landscape Conservation in Town Planning: The Case of Aya, Japan

The population of Aya is only 4000, but this case study is presented in this book for the following characteristics of Aya town:

- Unique town planning incorporating watershed protection, including forested areas.
- Urban development based on small-scale industries, ecotourism and organic farming practices. The case can also be important for rethinking the target scale of urbanization for sustainable water environments.
- Many of the urban areas in the world are either close to or inside protected areas, especially in developed parts of the world (Guneralp and Seto 2013), which compels us to think about how to incorporate the (still not well-integrated) biological parameters of water resource management into city and town developments in these areas.
- Aya puts importance on the protection of old growth forests. As old growth forests are resilient ecosystems, their conservation in or near urban domains can ensure a sustainable and stable water resource potential in cities.
- Aya represents newly urbanized area. In the coming decades, we will see increasing numbers of newly urbanized areas, many of which may be close to

Fig. 5.4 Location of the town of Aya in Kyushu, Japan (*Source* Google Map)

natural areas and protected areas. Therefore, the case of Aya is an example of a 'model' of what other urban areas should strive to be how an integrated landscape approach for water security and sustainability can be achieved.

The forests of Aya are located on a plateau area with nearby low mountains ranging from 200 to 600 m in height, characterized by deposits of volcanic ash called 'shirasu' and the alluvial plains of the Aya Kita and Aya Minami Rivers (Fig. 5.4). The Aya town has decided to integrate the biological landscape into town planning. This approach could not have been possible without strong political willingness coupled with an environmental regulator in the form of the Nature Conservation Society of Japan (NACS-J). Industry regulators are prominent bureaucratically, but long-term planning often does not work where economic returns (mainly counted in monetary terms) are low. Actions by land-use regulators and planners beyond the gridlock of industrial planning were possible due to the inclusion of NACS-J and political willingness (Shumiya et al. 2015).

The forests of Aya remain a natural showcase of broadleaved evergreen forests in Japan and Asia, but such forests only remain as fragmented remnants of their once widely distributed ecosystem, as the ecosystem is formed between a transition zone of tropical evergreen and temperate deciduous forests that originated as far back as the Cretaceous period (approximately 145 ma to 65 ma ago). These forests are called lucidophyllous forests. In Aya, 1033 botanical species, 145 fungal species, 70 avian species and 19 mammalian species have been documented (UNESCO 2012).

The warm-temperate evergreen broadleaf forest (WEBF) has a unique forest structure with distinctive canopy, lower tree, shrub and herb layers, all of which are composed of evergreen species (Miyawaki 1984). These forests are mainly dominated by species such as *Persea thunbergii, Castanopsis sieboldii, Lithocarpus edulis, Quercus (Cyclob.) salicina, Q. gilva, Distyliwn racemosutn, Casnnopsis cuspidata* and *Q. acuta* (Tagawa 1995). The smooth leaves of these forests efficiently reflect sunlight; Kira (1977) classified these forests as lucidophyll, referring to the shiny surfaces of the leaves. The richness of Aya's forest ecosystem contributes to the healthy and diverse Aya River and its two branches, the Aya Minami and Aya Kita, that provide for a range of different activities along the urban waterfront of the town of Aya. Aya's water has been named one of the 100 best forest springs, and the town has been named 'Water Town,' which encompasses multiple ecosystem benefits such as fresh air, sky watching, health, the beautification of urban areas through flower cultivation and organic and traditional agriculture.

The Aya BR is the result of a partnership between the Kyushu Regional Forest Office, Miyazaki Prefecture, Aya Town, the Nature Conservation Society of Japan (NACS-J) and a citizen group named the Teruha Forest Association for conserving the Aya WEBF. The Aya BR was recommended as a new BR site by the 23rd UNESCO Man and the Biosphere (MAB) Program and was added to the MAB in the next year. It is divided into seven zones for strategic management purposes: (1) wild nature, (2) the transition zone from wild nature to human domains, (3) agroecological areas, (4) farming areas, (5) residential areas, (6) areas with small-scale industries and (7) riverine areas.

The area covers parts of two cities (Kobayashi and Saito), two towns (all of Aya and part of Kunitomi) and part of one village (Nishimera) in the south-central part of Miyazaki Prefecture. The Aya BR has a core area of 682 ha with a (unusually[1]) large buffer area of 8,982 ha, and a transition area (village domain) of 4916 ha (Fig. 5.5). The area is the largest laurel forest in Japan with high species endemism, and these forests also connect to beech (*Fagus crenata*) forests at higher altitudes to village domains at lower elevations that practice organic agriculture. These are the major characteristics that have set Aya apart from other BRs; in particular, the inclusion of human domains (cities, towns and villages) in the conservation zones has been a new and unique component of this BR. In the past, human domains had logged off broadleaved forests across the Japanese Archipelago as well as throughout Asia, and this threat remains substantial at present. Both laurel and beech forests are noted for their complex water interception and retention characteristics (Figueira et al. 2013; Kirchen et al. 2017; Kudo 2015). It is from this perspective that the Aya BR requires special attention for maintaining healthy water ecosystems near town and urban areas.

[1]The small core area is a reminder of the dwindling laurel forest habitats, even in conservation areas.

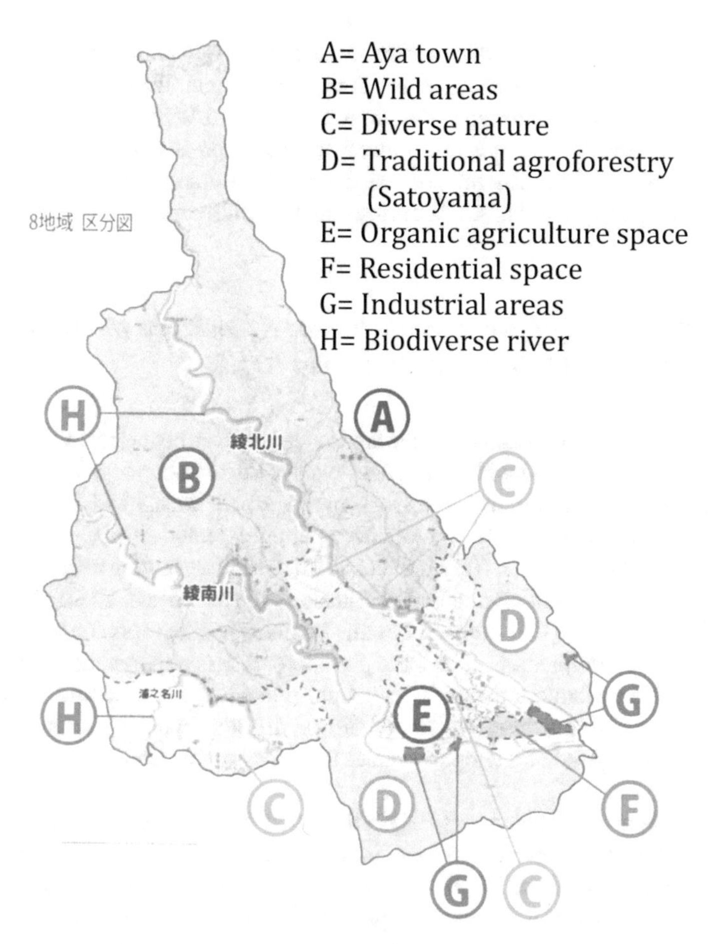

Fig. 5.5 Aya town planning map (above) with the visualization of diverse landscapes (below) (*Courtesy* Aya town office)

5.6.1.1 Challenges for Management

The Aya BR is now heavily fragmented. Despite the fact that Aya is an immensely important biosphere reserve and a nationally important ecological landscape in Japan, its future is threatened by a number of different factors. Some of the notable include environmental degradation due to concretized ditches and waterways, which do not allow 'riverscapes' to develop in the river and drainage valley, and are constructed with only water as a physical property. These structures also reduce the influence of rain and floodwaters in the river regime. Invasive species such as water hyacinth (*Eichhornia crassipes*) remain a common threat. This plant reduces the dissolved oxygen (DO) concentration in the water, affecting fish populations in Aya's rivers. The native tree population is affected by an increase in the number of shika deer (*Cervus nippon*), which consumes the saplings of native trees and impact forest regeneration. Industrial plantations of Japanese cypress (*Chamaecyparis obtuse*) and Japanese cedar (*Cryptomaria Japonica*) surround Aya and further isolate the native forest fragments with landscapes that do not connect to WEBF ecosystems. Overgrown and unused plantations are a threat to landscape sustainability as they not only reduce the local biodiversity, but the combined biomass can destabilize steeply sloped areas, prompting landslides during heavy rainfall events (Genet et al. 2008).

5.6.2 Prioritizing Diverse Stakeholder Knowledge for Water Resource Governance in Urban Areas

How diverse knowledge available at the landscape level can be integrated for effective environmental management is a challenge. Integration of knowledge has changed its scope, from knowledge transferred from experts to the locals to locals to experts (i.e., a primarily 'bottom up'), to ultimately mutual interaction or what is called as 'transdisciplinary' approach (Raymond et al. 2010) and a growing need to exchange knowledge among scientists and decision-makers (Cvitanovic et al. 2016).

Raymond et al.'s (2010) study argues that knowledge integration by academic and non-academic participants, utilization of participatory research methods and repeated processes of knowledge creation, application, reflection, learning and feedback to science or decision-making need to be built on different types on knowledges and across different spatial and temporal scales are necessary procedures for knowledge integration.

Local ecological knowledge (LEK) in this sense can be applied to capture a wide range of ecological knowledge present in the urban (and peri-urban) water environments (see also Chen et al. 2016). LEK can be used broadly also to capture different types of urban ecosystems from a highly modified and highly human influenced (i.e., city center, central business district where main natural components can be conserved in urban home gardens and agricultural spaces), to moderately human influenced (immediate peri-urban areas) to low human influenced (peri-urban areas that are far away from the urban center, protected areas near or inside an urban area). These represent complex ecosystems related to an urban area having different types of influences on the water environments. The East Kolkata Wetlands or (EKW)—a Ramsar site located near Kolkata conurbation in West Bengal state, India (Fig. 5.6)—is an example where the ecosystem is not highly modified and is connected to different set of local ecological knowledge that are connected with local livelihoods (i.e., artisanal fisheries, agriculture and horticulture with nutrients from wastewater influx) (Raychaudhuri et al. 2008; Bhattacharya et al. 2012). The local farmers and fishers maintain *bheries* (Fig. 5.7) or fish culture area and vegetable farming with the sewage from Kolkata as the main input. The locals maintain an ecological balance in the landscape through carefully managing the oxygen needs of the microorganisms, which cleanse the sewage, and the oxygen needs of the fish which feed on the microorganisms (Vicziany et al. 2017). However, with increasing industrial effluents, heavy

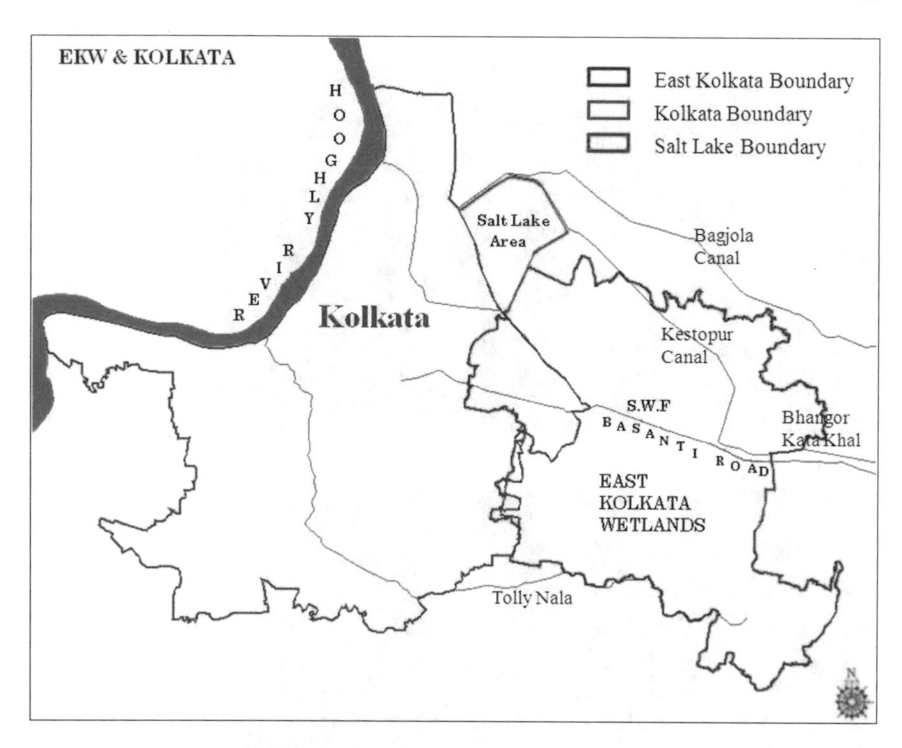

Fig. 5.6 Location of East Kolkata Wetland near Kolkata city (*Source* East Kolkata Wetland Management Authority, http://ekwma.in/ek/maps-2/)

metal pollution can seep in the wetland undetected by the LEK of the local farmers and fishers. Moreover, heavy metals are suspected to stay in the wetland system and can leach through to the underground aquifer and making the groundwater vulnerable to pollution (Sahu and Sikdar 2008). Therefore, the input of heavy metals whose presence cannot be detected by the LEK holders may transcend the ability of LEK systems and hinder the safe provision of food provisioning services from the EKW. It is argued that the one of the main pollutants, the tanneries, needs financial and compliance-related incentives to take actions not to inject untreated sewage into the EKW (Bagchi and Banerjee 2013). A reconsideration of knowledge partnership between scientific community and the local LEK holders is also a must for sustaining freshwater-related ES from the EKW. Considering the growing urbanization in Asia, ecosystems like that of EKW need more willingness of urban people to make this knowledge partnership possible to maintain and revive different ecosystem services that are strongly linked with provisioning (also seen in the Manila case study).

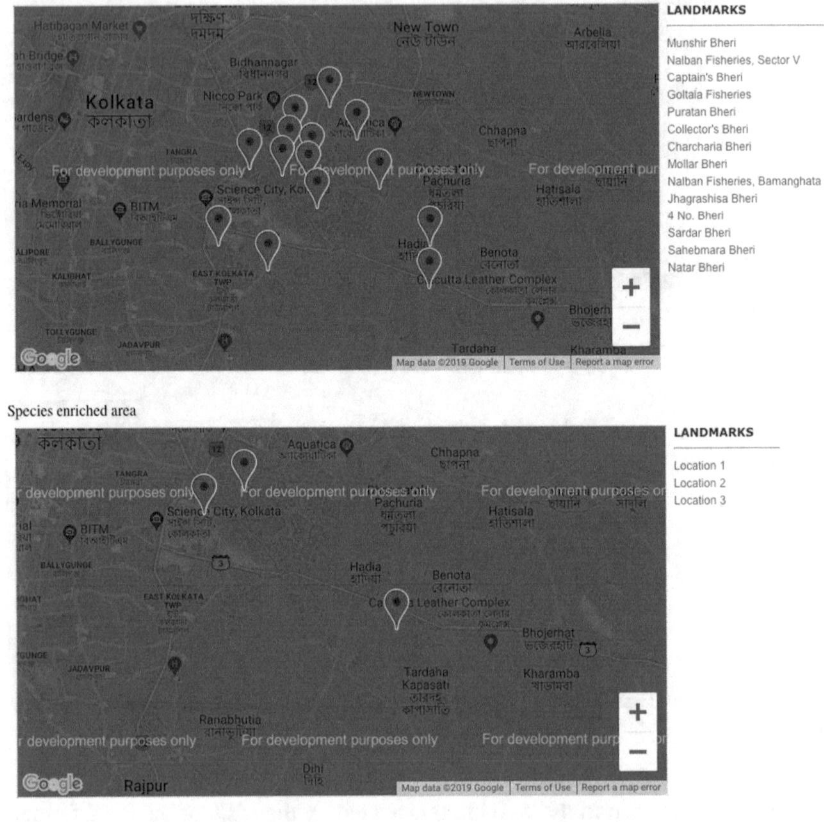

Fig. 5.7 Location of bheris (top) and species rich areas (bottom) within East Kolkata Wetland (*Source* East Kolkata Wetland Management Authority, http://ekwma.in/ck/maps-2/)

5.6.3 *Multiple Provisioning Services from Peri-Urban Watershed Environments: Case Study of Jala-Jala Watershed Near Manila*

The Manila conurbation is one of the largest metropolises in Asia, and it occupies part of the Pasig–Marikina–Laguna de Bay watershed, one of the most productive in the Philippines. Manila has also experienced regional urbanization that began as a venture to realize industrial growth through the Calabarzon Project, which was initiated to take industries outside Manila and into its peri-urban areas (Radford and Butardo-Toribio 1996; Kelly 2003). This type of growth has a background. The Philippines lagged behind in rice production when compared with their neighbors such as Thailand, Indonesia and Vietnam, and after the end of the Marcos administration and over 20 years of resource overexploitation, urbanization and industrialization seemed to be the only option for economic growth in the Philippines. However, the environments around Manila suffered as a result. The waters of peri-urban rivers such as the San Cristobal and Cabuyao became filthy and smelly with no fisheries in the downstream areas. The Marikina River also experienced the same fate, and it was only through the physical effort of cleaning the river that some sport fishing was made possible. Manila is also experiencing water shortage problems. The condition of the Laguna de Bay lake ecosystem has deteriorated due to pollution into the lake through its tributary rivers and waterways (Molina 2011). The freshwater lake has a high concentration of heavy metals in its western part. These heavy metals (arsenic, mercury, cadmium, chromium, copper, zinc and lead) find their ways by bioaccumulation into the bodies of freshwater fishes, such as milkfish (*Chanos chanos*), kanduli (*Neoarius berneyi*), bighead (*Hypophthalmichthys nobilis*), dalag (*Osmeriformes*) and tilapia (*Perciformes*).

Sustainable development for freshwater provision for household, sanitation and agricultural uses is a must for the Manila conurbation as it is the foundation of human well-being, without which natural environments and the associated benefits from ecosystems would also be in peril. Additionally, since the Philippines is a country under the influence of the El Nino and climate changes, forces that are far beyond strategic management interventions, the robustness of solutions to water-related issues lies in the resilience of its urban ecosystems that are related to water environments. These environments include rivers, lakes and watercourses, riparian areas, in-channel landscapes and landscapes between water courses. Hence, there is a need to incorporate these diverse landscapes to achieve better water resource management in the urban and peri-urban areas of Manila (Chakraborty et al. 2019).

From this perspective, an example is provided of the outskirts of the city of Manila. In 2016, a survey of 90 households was conducted in the peri-urban areas of the Manila conurbation (Fig. 5.8), and the results suggested that residents of peri-urban Jala-Jala area procure agricultural crops and non-marketed fruits, fish, a variety of medicinal plants and other non-timber forest products 'free' from the freshwater landscape (Fig. 5.9). These landscapes also have natural spring water in the valley as the main source for freshwater for drinking and irrigation. The Jala-Jala area represented examples of water-rich agroecological landscapes that also produce bundles of other ecosystem benefits near the peri-urban areas of a city, which is possible because of the variety of landscapes involved: secondary regenerative forests, croplands, grasslands, and open lands (but fewer built-up areas). This *diversity* of peri-urban landscapes presents the 'package,' where ES bundles (landscape diversity reflected in benefits) are available *along with* freshwater. The local residents of Jala-Jala had a number of different resource pool to support their livelihood and show a considerable amount of shared values toward these resources (Fig. 5.10). This suggests that conserving freshwater landscapes in peri-urban and areas of Manila has huge potential to release bundles of other associated ES benefits from the landscape.

These case studies show that landscapes of these peri-urban spaces hold the answer to maintaining diverse ecosystem benefits near urban areas. As it is difficult to newly create landscape elements such as urban home gardens or restored riparian and coastal areas with natural or semi-natural land covers that ensure better water retention

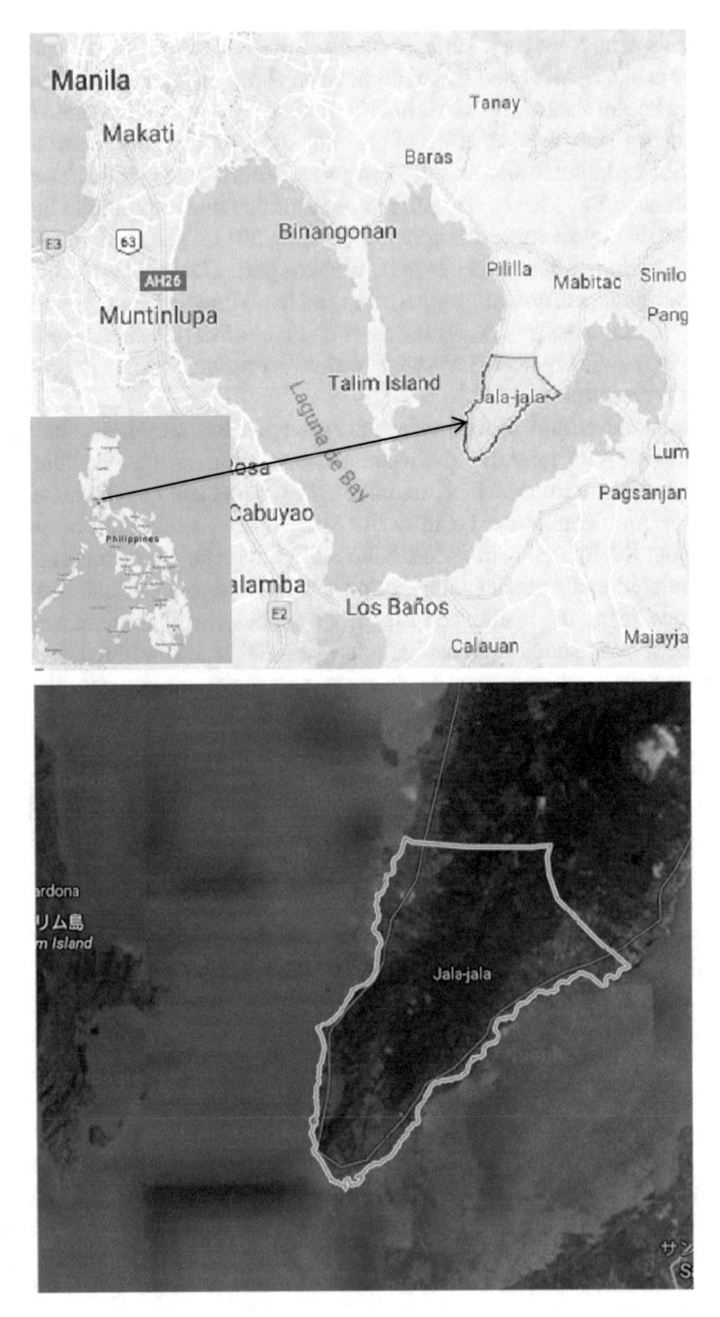

Fig. 5.8 Location of the Jala-Jala area in the Laguna de Bay Basin, the Philippines (*Source* Google Map)

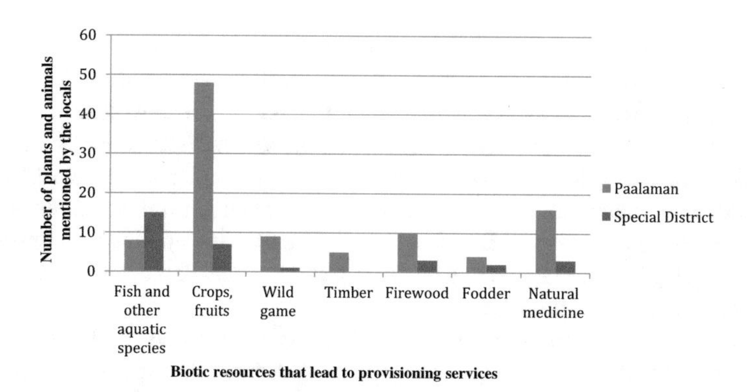

Fig. 5.9 Bundle of provisioning services acquired free from the landscape by the locals (*Source* Survey Data by the author)

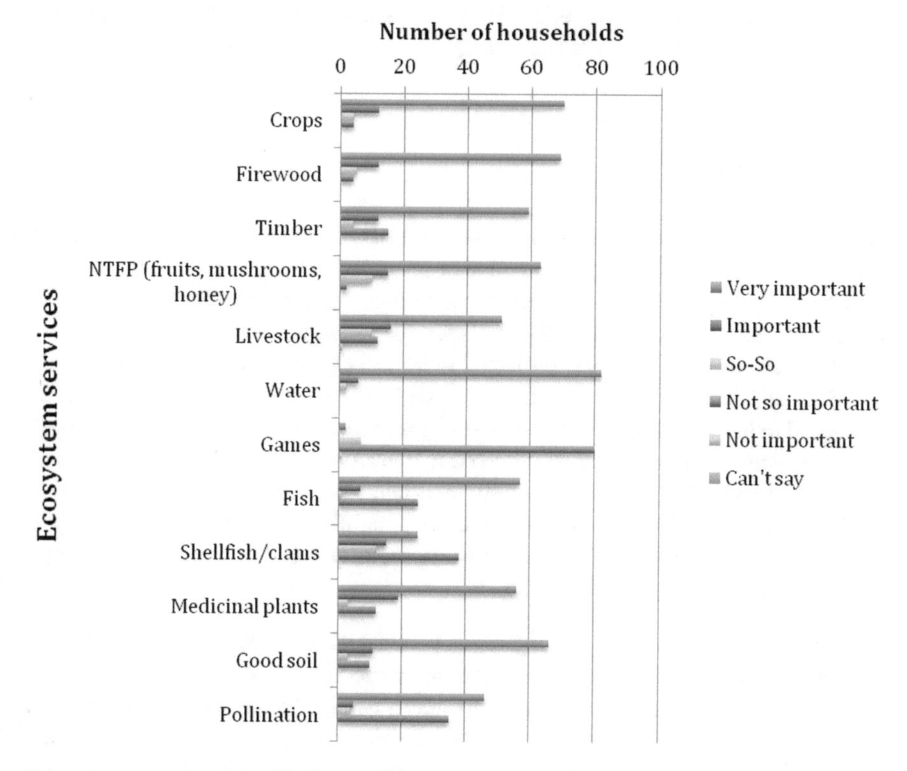

Fig. 5.10 Importance of different ecosystem services for each household in peri-urban Jala-Jala located in the outskirts of Manila ($N = 90$). The ecosystem services identified by the locals are highly interlinked and within the bundle of freshwater ecosystem services (*Source* Survey Data by the author)

and regulation (see also May 2006), the surrounding areas that contain natural and semi-natural areas remain key to retaining the multiple ecosystem service benefits associated with freshwater environments in urban areas. The case studies show that viable ways can be through maintaining remaining wild habitats (the case of Aya) or semi-natural areas by identifying and quantifying different ES available (the case of Jala-Jala) or by maintaining LEK systems that can maintain freshwater ecosystems near urban areas (the case of East Kolkata Wetland). This applies to a limited but rapidly growing body of knowledge on urban planning (Yli-Pelkonen and Kohl 2005; Chen et al. 2016; Chakraborty et al. 2019).

Urban rivers are the products of connectivity of upland and rural rivers. Efforts can also be toward redistributing the incentives generated through urban policy and decision-making and ensuring that local land-use practices continue in these areas. However, this approach is not the only solution; non-market-based instruments such as wisdom, morality and education to encourage an appreciation of nature as well as to appreciate the diverse benefits that watershed and riverine landscapes can provide to the urban and peri-urban areas, can play a vital role for their sustenance.

5.7 Summary

For a landscape-based approach to water security, there is a need to incorporate natural or semi-natural landscapes inside or near urban areas. This is because water environments depend on the landscape functions and connectivity. This becomes more important if we consider the statistics for urban areas near protected areas. By 2030, urban land uses near protected areas are expected to increase (Guneralp and Seto 2013). But urban areas distant from protected areas may not receive the direct and indirect benefits of diverse landscapes that can produce multiple ecosystem services. Hence, urban residents will need to conserve areas that maintain the basic ecosystem functions in order to get continuous benefits from these landscapes. Based on the arguments presented in this chapter, we highlight five important and highly interlinked measures that can increase water security together with other multiple benefits from the ecosystem in urban areas.

Supporting habitats Perhaps the greatest benefits are that healthy watershed landscapes can provide important habitats for many species. Urban areas lack biodiversity, so populations of urban areas cannot directly avail the many biodiversity-related services. The benefits include direct ones such as provision of food (without habitats, humans will not have any food for survival) or indirect ones such as pollination for agricultural practices or indirect ones such as cognitive development from experiencing urban green spaces and the associated biodiversity that enhances the mental and physical development of a person. Unaltered coastal areas can house sea grass and mudflats, in-channel pool riffle sequences and riparian habitats. These are fundamental biophysical characteristics of diverse streams and coastal areas. Urban gardens and home gardens can also provide secondary habitats for birds, insects and soil organisms in biodiversity-rich pockets that can be vital micro-habitats or buffer zones near protected areas that are near urban areas.

Natural sewage treatment Wetlands can restore over 90% of nitrogen and phosphorous (Ewel 1997), the two major water pollutants in the worlds rivers, lakes and coastal areas. Phytoremediation (a form of green technology) can be used more to remove certain contaminants from the water such as fluoride (Karmakar et al. 2016). Local ecological knowledge associated with time0tested methods of trial and error can prove a resilient way to treat urban sewage and extract multiple ecosystem benefits from otherwise polluted freshwater environments. Thus, the conservation and restoration of natural wetlands, including those which have rich local ecological knowledge base in the urban watersheds, can be viable best management practices (BMP), which are discussed in Chap. 6 of this volume.

Water supply Landscape conservation in and near urban areas can significantly secure the water supply. This chapter, together with other chapters of this volume (Chaps. 1, 5 and 6) suggests that landscapes can enhance the natural infiltration capacity and wastewater recycling capacity, helping groundwater recharge for reuse. The existence of cleaner water in a locality can produce other positive feedbacks such as food production, encouraging visitation to natural areas, education or the enjoyment of recreational spaces. A healthy river regime that supports different habitats (e.g., in-stream and riparian) ensures a better water network and ecological connectivity in urban areas, including a healthy groundwater regime that are needed to ensure sustained water supply.

Disaster risk reduction Many waterways, rivers and coastal areas are planned for reducing risks from disasters. Utilizing the ecosystem functions of diverse landscapes can be a low-key but high benefit-oriented approach for ecosystem-based disaster risk reduction in urban areas. A wide range of issues such as the reduction of monetary losses related to infrastructure, losses due to (mainly mono) crop failures and adverse impacts on crops can be achieved with ecosystem-based approaches (Renaud et al. 2013). Urban areas provide a sense of security, but this does not mean that they reduce environmental risks such as natural hazards (Mitchell 1999). The latest hurricane-induced flooding in the USA and monsoon-induced flooding in India suggest that storms and flooding can occur in despite man-made protection infrastructure and can cause considerable damage.

Spaces for recreation and cognitive interactions Urban water environments can produce recreational spaces for the cognitive development of children and adults, leading to good mental health and a balanced life. Urban areas produce stressful environments that are improper for child development, and recreational services are some of the most valued in urban areas (Bolund and Hunhammer 1997). People can also benefit from the stress-release functions from these places; their aesthetic beauty can attract tourists and can be sources of direct income for the managers.

We have seen that addressing water security issues by involving the whole landscape can be a viable solution to most urban water security problems. Moreover, such an approach can bring multiple ecosystem benefits from the landscape. One important consideration is that these ecosystem services should not be seen as separate entities, and efforts to avail them should not be confined to reductionist approaches that deal with only one part of the ecosystem (only riparian areas or only maintaining pool riffle sequences in the case of riverscape management). The connectivity of functions across urban ecosystems should be kept in mind to ensure the internalization of multiple and invisible ecosystem values that are mostly disregarded in urban areas.

References

Adhikari, Asti, Noro (2010) Flood-related disaster vulnerability: an impending crisis of megacities in Asia. J Flood Risk Manag 3(3):185–191. https://doi.org/10.1111/j.1753-318X.2010.01068.x

Ansar A, Flyvbjerg B, Budzier A, Lunn D (2014) Should we build more large dams? The actual costs of hydropower megaproject development. Energy Policy 69:43–56. https://doi.org/10.1016/j.enpol.2013.10.069

Bagchi S, Banerjee S (2013) Efficient pollution management through CETP: the case of the calcutta leather complex. In: Banerjee S, Chakrabarti A (eds) Development and sustainability: India in a global perspective. Springer, New Delhi, p 523

BBC (2017) How the demand for sand is killing rivers. http://www.bbc.com/news/magazine-41123284

Berkowitz AR, Nilon CH, Hollweg KS (eds) (2003) Understanding urban ecosystems: a new frontier for science and education. Springer, New York

Bhattacharya S, Ganguli A, Bose S, Mukhopadhyay A (2012) Biodiversity, traditional practices and sustainability issues of East Kolkata Wetlands: A significant Ramsar site of West Bengal (India). Res Rev Biosci 6:340–347

Boken VK (2016) Groundwater crisis of a mega city: a case study of New Delhi, India. In: Maheswari B, Singh VP, Thoradeniya B (eds) Balanced urban development: options and strategies for liveable cities, pp 211–219. Springer

Bolund P, Hunhammer S (1997) Ecosystem services in urban areas. Ecol Econ 29(2):293–301

Bon HD, Parrot L, Maustier P (2010) Sustainable urban agriculture in developing countries. A Rev Agron Sustain Dev 30:21–32

Bonan GB (2002) Ecological climatology: concepts and applications. Cambridge University Press, New York, NY

Bowler DE, Buyung-Ali L, Knight TM, Pullin A (2010) Urban greening to cool towns and cities: a systematic review of the empirical evidence. Landscape Urban Plan 97(3):147–155

Bunn SE, Arthington AH (2002) Basic principles and ecological consequences of altered flow regimes for aquatic biodiversity. Environ Manage 30(4):492–507

CBD (2010) COP decision X/2. Strategic plan for biodiversity 2011–2020. Available at: http://www.cbd.int/decision/cop/?id=12268. Accessed 1 Oct 2013

Chakraborty S, Avtar R, Raj R, Minh HVT (2019) Village level provisioning ecosystem services and their values to local communities in the peri-urban areas of Manila, The Philippines. Land (MDPI) 8(177):1–18. https://doi.org/10.3390/land8120177

Chakraborty S, Chakraborty A (2017) Satoyama landscapes and their change in a river basin context: lessons for sustainability. Issues Soc Sci 5(1):38–64

Chan KA et al (2012a) Rethinking ecosystem services to better address and navigate cultural values. Ecol Econ 74:8–18

Chan KMA et al (2012b) Where are cultural and social in ecosystem services? A Framework for Constructive Engagement. BioScience 62(8):744–756

Chen C, Meruk H, Cheng et al (2016) Incorporating local ecological knowledge into urban riparian restoration in a mountainous region of Southwest China. Urban Forest Urban Greening 20:140–151. https://doi.org/10.1016/j.ufug.2016.08.013

Chichilnisky G, Heal G (1998) Economic returns from the biosphere. Nature 391:629–630. https://doi.org/10.1038/35481

Chiesura A (2004) The role of urban parks for the sustainability of cities. In Marchettini N, Brebbia CA, Tiezzi E, Wadhwa LC (eds) The sustainable city III, Accessed from https://www.witpress.com/Secure/elibrary/papers/SC04/SC04034FU.pdf

Conway H (2000) Parks and people: The social functions. In: Woudstra J, Fieldhouse K (eds) The Regeneration of Public Parks, E&FN Spon, London, UK

Costanza R, d'Arge R, deGroot R, Farber S, Grasso M, Hannon B, Limburg K, Naeem S, O'Neill RV, Paruelo J, Raskin RG, Sutton P, van den Belt M (1997) The value of world's ecosystem services and natural capital. Nature 387:253–260

Cvitanovic C, Mcdonald J, Hobday A (2016) From science to action: principles for undertaking environmental research that enables knowledge exchange and evidence-based decision-making. J Environ Manag 183:864–874. https://doi.org/10.1016/j.jenvman.2016.09.038

Dadvand P, Nieuwenhuijsen MJ, Esnaola M, Forns J, Basagana X, Alvarez-Pedrerol M et al (2015) Green spaces and cognitive development in primary schoolchildren. PNAS 112(26):7937–7942. https://doi.org/10.1073/pnas.1503402112

Diaz S, Demissew S, Carabias J et al (2015) The IPBES conceptual framework—connecting nature and people. Curr Opin Environ Sustain 14:1–16. https://doi.org/10.1016/j.cosust.2014.11.002

Duraiappah AK, Nakamura K, Takeuchi K, Watanabe M, Nishi M (2012) Satoyama-satoumi ecosystems and human well being: socio-ecological production landscapes of Japan. United Nations University Press, Tokyo

Elmi A, Madramootoo C, Egeh M, Liu A, Hamel C (2002) Environmental and agronomic implications of water table and nitrogen fertilization management. J Environ Qual 31(6):1858–1867

Escobedo FJ, Kroeger T, Wagner JE (2011) Urban forests and pollution mitigation: analyzing ecosystem services and disservices. Environ Pollut 159(8):2078–2087

Esther S, Devadas MD (2016) A calamity of a severe nature: case study—Chennai, India. In: Proverbs D, Mambretti S, Brebbia CA, Ursino N (eds) Urban water systems and floods. WIT Press, Ashhurst, Southampton

Everard M, Moggridge HL (2012) Rediscovering the value of urban rivers. Urban Ecosyst 15:293–314. https://doi.org/10.1007/s11252-011-0174-7

Ewel KC (1997) Water quality improvement by wetlands. In: Daily GC (ed) Natures services. Societal dependence on natural ecosystems. Island Press, Washington, DC, pp 329–344

Figueira C, Menezes de Sequeira M, Vasconcelos R, Prada S (2013) Cloud water interception in the temperate laurel forest of Madeira Island. Hydrol Sci J 58(1):152–161

Fitzhugh TW, Richter BD (2004) Quenching urban thirst: growing cities and their impacts on freshwater ecosystems. Bioscience 54:741–754

Fujiwara S (2009) "Midori no Damu" no Hozoku: Nihon no Shinrin wo Ureu [Preserving "green dams": on supporting forests of Japan]. Ryokufu Shuppan, Tokyo, Japan

Garcia X, Barceló D, Comas J, Corominas L, Hadjimichael A, Page TJ, Acuña V (2016) Placing ecosystem services at the heart of urban water systems management. Sci Total Environ. https://doi.org/10.1016/j.scitotenv.2016.05.010

Genet M, Kokutse N, Stokes A, Foucard T, Cai X, Ji J, Mickovski S (2008) Root reinforcement in plantations of *Cryptomeria japonica* D. Don: effect of tree age and stand structure on slope stability. Forest Ecol Manag 256(8):1517–1526

Gómez-Baggethun E, Gren A, Barton DN, Langemeyer J, McPhearson T, O'Farrell P, Andersson E, Hamstead Z, Kremer P (2013) Urban ecosystem services. In: Elmqvist T et al (eds) Urbanization, biodiversity and ecosystem services: challenges and opportunities. Springer, Dordrecht

Grimm NB, Faeth SH, Golubiewski NE, Redman CL, Wu J, Bai X, Briggs JM (2008) Global change and the ecology of cities. Science 319:756–760

Guneralp B, Seto KC (2008) Environmental impacts of urban growth from an integrated dynamic perspective: A case study of Shenzhen South China. Glob Environ Change 18(4):720–735
Guneralp B, Seto KC (2013) Futures of global urban expansion: uncertainties and implications for biodiversity conservation. Environ Res Lett 8(1). Accessed from http://iopscience.iop.org/article/10.1088/1748-9326/8/1/014025
Gurnell A, Lee M, Souch C (2007) Urban rivers: hydrology, geomorphology, ecology and opportunities for change. Geography Compass 1(5):1118–1137. https://doi.org/10.1111/j.1749-8198.2007.00058.x
Haines-Young R, Potschin M (2011) Common international classification of ecosystem services (Update). Paper prepared for discussion at the expert meeting on ecosystem accounts organised by the UNSD, the EEA and the World Bank, London. European Environmental Agency
Haq AMS (2011) Urban green spaces and an integrative approach to sustainable environment. J Environ Prot 2:601–608. https://doi.org/10.4236/jep.2011.25069
Haslam SM (2008) The riverscape and the river. Cambridge
Hopwood JL (2008) The contribution of roadside grassland restorations to native bee conservation. Biol Cons 141(10):2632–2640
Huber CV (1989) A concerted effort for water quality. J Water Pollut Control Fed 60:483–496
IPBES (n.d.) Intergovernmental panel form biodiversity and ecosystem services. http://ipbes.net
Jenerette GD, Harlan SL, Stefanov WL, Martin CA (2011) Ecosystem services and urban heat riskscape moderation: water, green spaces, and social inequality in Phoenix, USA. vol 21, Issue 7, pp 2637–2651. https://doi.org/10.1890/10-1493.1
Jennings V, Larson L, Yun J (2016) Advancing sustainability through urban green space: cultural ecosystem services, equity, and social determinants of health. Int J Environ Res Public Health 13(2):196. https://doi.org/10.3390/ijerph13020196
Kamiyama C, Hashimoto S, Kohsaka R, Saito O (2016) Non-market food provisioning services via homegardens and communal sharing in satoyama socio-ecological production landscapes on Japan's Noto peninsula. Ecosyst Serv 17:185–196
Karelva et al (eds) (2011) Natural capital: theory and practice of mapping ecosystem services. Oxford University Press
Karmakar S, Mukherjee J, Mukherjee S (2016) Removal of fluoride contamination in water by three aquatic plants. Int J Phytorem 18(3):222–227. https://doi.org/10.1080/15226514.2015.1073676
Karr J (1991) biological integrity: a long-neglected aspect of water resource management. Ecol Appl 1(1):66–84
Kelly P (2003) Urbanization and politics of land in the manıla region. Ann Am Acad Polit Soc Sci 90:170–187. https://doi.org/10.1177/0002716203256729
Khatri N, Tyagi S (2014) Influences of natural and anthropogenic factors on surface and groundwater quality in rural and urban areas. Front Life Sci 8(1). https://doi.org/10.1080/21553769.2014.933716
Kira T (1977) A climatological interpretation of Japanese vegetation zones. In: Miyawaki A, Tuxen R (eds) Vegetation Science and Environmental Protection pp 21–30, Maruzen, Tokyo
Kirchen G, Calvarusi C, Granier A, Redon P, van der Heijden G, Breda N, Turpault MP (2017) Local soil type variability controls the water budget and stand productivity in a beech forest. For Ecol Manage 390:89–103
Kudo A (2015) Shirakami Sanchi wa Midori no Damu ni Narieruka: 522 Shirakami Sanchi ga Karyu Kasen ni Oyobosu Eikyo (Can 523 Shirakami Mountains act as a green dam?: The effect of Shirakami 524 forests on downstream drainage). In: The Shirakami Institute for 525 Environmental Sciences, Hirosaki University (ed) Shirakamigaku 526 Nyumon (Introduction to the study of Shirakami Mountains). 527 Hirosaki University Publication, Hirosaki, pp 60–65
Kummu M, De Moel H, Ward PJ, Varis O, Perc M (2011) How close do we live to water? A global analysis of population distance to freshwater bodies. PLoS ONE 6(6):e20578
Lankao PR, Qin H (2011) Conceptualizing urban vulnerability to global climate and environmental change. Curr Opin Environ Sustain 3(3):142–149

Lavigne F, Wassmer P, Gomez C, Davies TA, Hadmoko DS, Iskandarsyah TYWM, Gaillard JC, Fort M, Texier P, Heng MB, Pratomo I (2014) The 21 February 2005, catastrophic waste avalanche at Leuwigajah dumpsite, Bandung, Indonesia. Geoenviron Disasters 1:10. https://doi.org/10.1186/s40677-014-0010-5

Lee ACK, Maheswaran R (2011) The health benefits of urban green spaces: a review of the evidenc. J Public Health (Oxf) 33(2) 212–222 https://doi.org/10.1093/pubmed/fdq068

Lempérière F (2017) Dams and Floods. Engineering 3(1):144–149. https://doi.org/10.1016/J.ENG.2017.01.018

Levin S (1998) Ecosystems and the biosphere as complex adaptive systems. Ecosystems 1:431–436. https://doi.org/10.1007/s100219900037

Livesley SJ, McPherson GM, Calfapietra C (2016) The urban forest and ecosystem services: impacts on urban water, heat, and pollution cycles at the tree, street, and city scale. J Environ Qual 45(1):119. https://doi.org/10.2134/jeq2015.11.0567

Long H, Liu Y, Hou X, Li T, Li Y (2014) Effects of land use transitions due to rapid urbanization on ecosystem services: implications for urban planning in the new developing area of China. Habitat Int 44:536–544. https://doi.org/10.1016/j.habitatint.2014.10.011

Luttik J (2000) The value of trees, water and open space as reflected by house prices in the Netherlands. Landscape Urban Plan 48:161–167

Martin-Lopez B, Iniesta-Arandia I, Garcia-Llorente M, Palomo I, Casado-Arzuaga I et al (2012) Uncovering ecosystem service bundles through social preferences. PLoS ONE 7(6):e38970. https://doi.org/10.1371/journal.pone.0038970

May R (2006) "Connectivity" in urban rivers: conflict and convergence between ecology and design. Technol Soc 28(4):477–488. https://doi.org/10.1016/j.techsoc.2006.09.004

McArthur R (1955) Fluctuations of animal populations and a measure of community stability. Ecology. https://doi.org/10.2307/1929601

McCormack G (2007) Modernity, water, and the environment in Japan. In: Tsutsui WM (ed) A companion to Japanese history. Blackwell, Malden, MA, pp 443–459

McDonald RI, Forman RTT, Kareiva P et al (2009) Urban effects, distance, and protected areas in an urbanizing world. Landscape Urban Plann 93(1):63–75. https://doi.org/10.1016/j.landurbplan.2009.06.002

McDonald RI, Weber K, Padowski J, Florke M, Schneider C, Green PA, Gleeson T, Eckman S, Lehner B, Balk D, Boucher T, Grill G, Montgomery M (2014) Water on an urban planet: urbanization and the reach of urban water infrastructure. Glob Environ Change 27:96–105. https://doi.org/10.1016/j.gloenvcha.2014.04.022

MEA (Millennium Ecosystem Assessment) (2005) Ecosystems and humen well-being, synthesis. Island Press, Washington DC

Merry SM, Kavazanjian E, Fritz WU (2005) Reconnaissance of the July 10, 2000, Payatas landfill failure. J Perform Constructing Facil 19(2):100–107. https://doi.org/10.1061/(ASCE)0887-3828(2005)19:2(100)

Millennium Ecosystem Assessment (MA) (2005) Ecosystems and human well-being: current state and trends: findings of the Condition and Trends Working Group. Island Press, Washington, Covelo, London

Milliken (2018) Chapter 1.2—ecosystem services in urban environments. Nat Based Strat Urban Build Sustain 17–27 https://doi.org/10.1016/B978-0-12-812150-4.00002-1

Mitchell JK (1999) Crucibles of hazard: mega cities and disasters in transition. United Nations University Press, Tokyo

Miyawaki A (1984) A Vegetation ecological view of the Japanese Archipelago. Bull Inst Environ Sci Technol Yokohama Natn Univ 11:85–101

Molina VB (2011) Health risk assessment of heavy metals bioaccumulation in Laguna de Bay fish products. In: 14th world lake conference, Austin Texas. Retrieved from: https://www.pref.ibaraki.jp/soshiki/seikatsukankyo/kasumigauraesc/04_kenkyu/kaigi/documents/kosyou/14/2011wlc_victoriob.molina.pdf

Morton LW, Bitto EA, Oakland MJ, Sand M (2008) Accessing food resources: rural and urban patterns of giving and getting food. Agric Hum Values 25(1):107–119

Moseley L (2005) Water quality of rainwater harvesting systems (SOPAC Miscellaneous Report No. 579. Accessed from: http://www.pacificwater.org/userfiles/file/MR0579.pdf

Nazemi A, Madani K (2018) Urban water security: emerging discussion and remaining challenges. Sustain Cities Soc 41:925–928. https://doi.org/10.1016/j.scs.2017.09.011

Niemela J, Saarela S, Soderman T, Kopperoinen L, Yli-Pelkonen V, Väre S, Johan Kotze DJ (2010) Using the ecosystem services approach for better planning and conservation of urban green spaces: a Finland case study. Biodivers Conserv 19(11):3225–3243

O'Brien A, Townsend K, Hale R, Sharley D, Pettigrove V (2016) How is ecosystem health defined and measured? A critical review of freshwater and estuarine studies. Ecol Ind 8:722–729. https://doi.org/10.1016/j.ecolind.2016.05.004

Oberndorfer E, Lundholm J, Bass B, Coffman RR, Doshi H, Dunnett N, Gaffin S, Köhler M, Karen K, Liu Y, Rowe B (2007) Green roofs as urban ecosystems: ecological structures, functions, and services. Bioscience 57(10):823–833. https://doi.org/10.1641/B571005

Odum EP (1953) Fundamentals of ecology. W. B. Saunders Company, Philadelphia

Orsini F, Gasperi D, Marchetti L, Piovene C, Draghetti S, Ramazzotti S (2014) Exploring the production capacity of rooftop gardens (RTGs) in urban agriculture: the potential impact on food and nutrition security, biodiversity and other ecosystem services in the city of Bologna. Food Secur 6(6):781–792. https://doi.org/10.1007/s12571-014-0389-6

Padmalal D, Maya K, Sreebha S, Sreeja R (2008) Environmental effects of river sand mining: a case from the river catchments of Vembanad lake, Southwest coast of India. Environ Geol (54(4):879–889

Panyadee P, Balslev H, Wangpakapattanawong P, Inta A (2016) Woody plant diversity in urban homegardens in northern Thailand. Econ Bot 70(3):285–302

Pascual et al (2017) Valuing nature's contributions to people: the IPBES approach. Curr Opinion in Environment and Sustainability 26:7–16

Petts J and Gray AJ (2006) SMURF and the public: engagement and learning Report of EU Life Project—LIFE02 ENV/UK/ 000144 Environment Agency, Bristol

Postel S, Carpenter S (1997) Freshwater ecosystem service. In: Daily GC (ed) Nature's services: societal dependence on natural ecosystems. Island Press, Washington D. C., pp 195–214

Radford GJ, Butardo-Toribio M (1996) Environmental planning and management concerns in the Calabrzon area, Philippines. Environment and Resource Management Project (ERMP), Philippines. ERMP report 30

Rahman A et al (2016) Impact of urban expansion on farmlands: a silent disaster. In: Urban disasters and resilience in Asia, pp 91–112. https://doi.org/10.1016/B978-0-12-802169-9.00007-0

Raychaudhuri S, Misra M, Nandy P, Thakur AR (2008) Waste management: A case study of ongoing traditional practices at East Calcutta Wetland. J Am Agri Biol Sci 3(1). https://doi.org/10.3844/ajabssp.2008.315.320

Raymond CM, Fazey I, Reed MS, Stringer LC, Robinson GM, Eveley AC (2010) Integrating local and scientific knowledge for environmental management. J Environ Manag 91:1766–1777

Renaud FG, Sudmeier-Rieux K, Estrella M (2013) The role of ecosystems in disaster risk reduction. United Nations University Press, Tokyo

Saarikoski H, Primmer E, Saarela SR et al (2018) Institutional challenges in putting ecosystem service knowledge in practice. Ecosyst Serv 29 C:579–598. https://doi.org/10.1016/j.ecoser.2017.07.019

Sahu P, Sikdar PK (2008) Hydrochemical framework of the Aquifer in and around East Kolkata Wetlands, West Bengal, India. Environ Geol 55:823–835

Sala M (1992) Some hydrologic effects of urbanization in Catalan rivers. Catena 19(3–4):363–378. https://doi.org/10.1016/0341-8162(92)90009-Z

Schmidt C, Krauth T, Klöckner P, Römer MS, Stier B, Reemtsma T, Wagner S (2017) Estimation of global plastic loads delivered by rivers into the sea. Geophys Res Abs 19, EGU2017–12171, EGU General Assembly

Schneider A, Mertes CM, Tatem AJ, Tan B, Sulla-Menashe D, Graves SJ, … Dastur A (2015) A new urban landscape in East-Southeast Asia, 2000–2010. Environ Res Lett 10(3) https://doi.org/10.1088/1748-9326/10/3/034002

Schröter M, van der Zanden EH, van Oudenhoven APE, Remme RP, Serna-Chavez HM, de Groot RS, Opdam P (2014) Ecosystem services as a contested concept: a synthesis of critique and counter-arguments. Conserv Lett 7:514–523

Sharma SS, Chhabra SK (2015) Understanding the chemical metamorphosis of Yamuna River due to pollution load and Human use. Int Res J Environ Sci 4(2):58–63

Shumiya T, Okonogi H, Kawano K, Ishida T, Soma M (2015) Conservation of endangered ecosystem of warm-temperate evergreen broad-leaved forest (lucidophyllus forest) with local community: Implementation of collaborative management in Aya Town, southern Kyushu, Japan. Jpn J Conserv Ecol 18(2):225–238. https://doi.org/10.18960/hozen.18.2_225

Siciliano G, Urban F, Kim S, Lonn PD (2017) Hydropower, social priorities and the rural–urban development divide: The case of large dams in Cambodia. Energy Policy 86:273–285. https://doi.org/10.1016/j.enpol.2015.07.009

Simmons DL, Raynolds RJ (1982) Effects of urbanization on base flow of selected south shore streams, Long Island New York. J Am Water Resour Assoc 18(5):797–805

Singh NP (2010) Space and groundwater problem in Delhi. Proc Environ Sci 2:407–415. https://doi.org/10.1016/j.proenv.2010.10.045

Su GS (2005) Water-borne illness from contaminated drinking water sources in close proximity to a dumpsite in Payatas, The Philippines. J Rural Trop Public Health 4:43–48

Tagawa H (1995) Distribution of lucidophyll oak laurel forest formation in Asia and other areas. Tropics 5(1/2):1–40

Takeuchi K, Brown RD, Washitani I, Tsunekawa A, Yokohari M (eds) (2003) Satoyama: the traditional rural landscape of Japan. Springer, New York

TEEB (2010) The economics of ecosystems and biodiversity. Ecological and Economics Foundations, Earthscan, London

The World Bank (2009) Poverty and social impact analysis of groundwater over-exploitation in Mexico. http://siteresources.worldbank.org/INTPSIA/Resources/490023-1120841262639/Mexico_groundwater.pdf

UN (2012) World urbanization prospects: The 2011 revision highlights. United Nations Department of Economic and Social Affairs/Population Division, New York http://esa.un.org/unup/pdf/WUP2011_Highlights.pdf

UNESCO (2012) Ecological sciences for sustainable development: Aya. http://www.unesco.org/new/en/natural-sciences/environment/ecological-sciences/biosphere-reserves/asia-and-the-pacific/japan/Aya/

Vicziany M, Chattopadhyay D, Bhattacharyya S (2017) Food from sewage: fish from the East Kolkata wetlands and the limits of traditional knowledge. South Asia: J South Asian Stud. http://doi.org/10.1080/00856401.2017.1341038

Waley P (2000) Following the flow of Japan's river culture. Japan Forum 12(2):199–217

Wiens JA (2002) Riverine landscapes: Taking landscape ecology into the water. Freshw Biol 47(4):501–515. https://doi.org/10.1046/j.1365-2427.2002.00887.x, http://onlinelibrary.wiley.com/doi/10.1046/j.1365-2427.2002.00887.x/full

Wolch JR, Byrne J, Newell, JP (2014) Urban green space, public health, and environmental justice: the challenge of making cities 'just green enough', vol 125, May 2014, pp 234–244. https://doi.org/10.1016/j.landurbplan.2014.01.017

Yli-Pelkonen V, Kohl J (2005) The role of local ecological knowledge in sustainable urban planning: perspectives from Finland, sustainability: science. Pract Policy 1(1):3–14. https://doi.org/10.1080/15487733.2005.11907960

Chapter 6
Urban Stormwater Management: Practices and Governance

6.1 Background

With rapid global changes in past few decades, it is very important for us to visualize the different perspectives of global challenges and its effect on different natural resources in particular water resources. With huge gap between demand and supply along with spatiotemporal disparity, water resources especially in urban pockets around the world are either approaching or exceeding the limits of sustainable use at frightening rates. With rapid expansion of concrete jungle and high groundwater extraction rate, we are encountering fast groundwater quality and quantity deterioration and increasing flood events. Therefore, it is very important to improve existing water management systems for restoring high-quality water and reducing the frequency of hydrometeorological disasters while preserving our natural/pristine environment in a sustainable manner. To build a robust water management system, one of the important tasks is optimal collection, infiltration and storage of stormwater. In general, stormwater runoff is rainfall that flows over the ground surface, large volumes of which are quickly transported to local water bodies and can cause flooding and coastal erosion as well as potentially carry different pollutants on the paved way to the water bodies. Henceforth, this chapter strives to explore optimal capture measures for the sustainable stormwater management. This chapter also tries to explain the vital role remote sensing and GIS technology can play in designing optimal capture measures under the threat of future extreme weather events and climate change. Apart from technological and structural components, this chapter also tries to capture community attitudes, which are influenced by a range of factors, including knowledge

---

DISCLAIMER: This chapter is adopted from the article first published in Environmental Science and Policy by Saraswat et al. 2016 whose reference is as follows: C. Saraswat, P. Kumar, B.K. Mishra. Assessment of stormwater runoff management practices and governance under climate change and urbanization: an analysis of Bangkok, Hanoi and Tokyo Environ. Sci. Pollut., 64 (2016), pp. 101–117, https://doi.org/10.1016/j.envsci.2016.06.018.

B. K. Mishra et al., *Sustainable Solutions for Urban Water Security*, Water Science and Technology Library 93,
https://doi.org/10.1007/978-3-030-53110-2_6

of urban water problems, are also considered to assess stormwater runoff management practices for achieving urban water security. In order to consider, abovementioned factors, this chapter explored different characteristics of stormwater runoff management policies and strategies adopted by three different case studies in Asia, namely Japan, Vietnam and Thailand. Finally, it presents comparative measures to manage water scarcity and achieve water resiliency, as well as presents an overview of stormwater management to guide future optimal stormwater measures and management policies within the governance structure. Furthermore, the effectiveness of different on-site facilities, including those for water harvesting, reuse, ponds and infiltration, is explored to highlight best feasible adaptation strategies that restore the water cycle and reduce climate change-induced flooding and water scarcity on a catchment scale.

6.2 Global Overview

The global population has reached 7.2 billion, and more people now live in cities than in rural areas (UNDESA 2014). Water is a very critical natural resource for the world's fastest growing urban areas, and commercial, residential and industrial users already place considerable demands on the water resources and supply of cities; therefore, water treatment is often required (Bahri 2012). The demand for water resources in urban areas is approaching the water supply capacity, and in many cases, the limits of sustainable water use are being exceeded (Hatt et al. 2004; Mitchell et al. 2003). Water scarcity sometimes leads to conflicts over water rights, and in urban watersheds, competition with agriculture and industry is intensifying as cities expand in size and political influence (Bahri 2012). With industrial and domestic water demand expected to double by 2050 (UNDP 2006), competition among urban, peri-urban and rural areas will likely worsen. Therefore, a critical challenge for newly developed urban cities is to design for resilience to the impacts of climate change with regard to sustainable management of water resources. It is currently well accepted that the conventional urban water management approach is highly unsuited to addressing current and future sustainability issues especially under the under global changes and extreme weather conditions (Ashley et al. 2005; Wong and Brown 2008). The conventional approach to urban water systems around the world involves the use of a similar series of systems for storm water drainage, potable water and sewerage, but as explained by Bahri (2012), the unsustainable nature of this approach is highlighted by the current ecosystem-related problems and degraded environment in urban areas due to changes in the hydrology of catchments and the quality of runoff, leading to modified riparian ecosystems (Bahri 2012). United Nations Agenda 21 (United Nations Agenda 21 1992) stated that achieving sustainable urban water systems and protecting the quality and quantity of freshwater resources are key components of ecologically sustainable development. Because of climate change and the spread of urbanization, the negative impacts on water resources are intensifying, resulting in increasing runoff, pollutant loads and pressure on existing systems, but augmenting

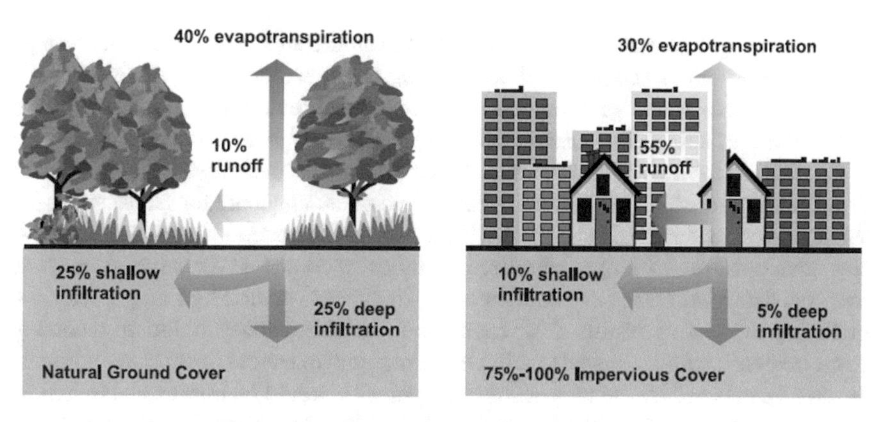

Fig. 6.1 Comparison between natural ground cover and >75% impervious land cover and its effect deep infiltration and surface runoff. (*Source* FISRWG 1998)

conventional systems will require a significant economic cost. Therefore, alternative approaches are required to develop sustainable water systems in urban environments, and integrated urban water management (IUWM) is one such approach that views the water supply, drainage and sanitation as components of an integrated physical system within an organizational and natural landscape (Mitchell et al. 2007). IUWM is an integrated system that seeks to reduce the inputs and outputs to decrease the inefficiencies of water resource systems that are associated with traditional urbanization practices (Hardy et al. 2005). Although this incorporation and diversification of urban water systems increase the complexity of urban water systems, they also provide more opportunities to attain sustainable water use and increase the overall resilience of the water system (Mitchell and Diaper 2005; Mitchell et al. 2007). The key components of the IUWM system are the methods and measures to capture and utilize urban storm water, pollution control, groundwater recharge, etc. Figure 6.1 defines stormwater as precipitation, such as rain or melting snow, and in a natural environment, a small percentage of precipitation becomes surface runoff. However, as urbanization increases, the amount of surface runoff drastically increases because of surface concretization. Surface runoff is created when pervious or impervious surfaces are saturated from precipitation or snow melt (Durrans 2003). Pervious surface areas naturally absorb water to the saturation point, after which rainwater becomes runoff and travels via gravity to the nearest stream. This saturation point is dependent on the landscape, soil type, evapotranspiration and biodiversity of the area (Pierpont 2008).

Due to the impervious surfaces that cover the natural environment in an urban setting, the hydrological processes of surface water runoff become more unnatural, damaging infrastructure and contaminating water with pollutants (Ragab et el. 2003). The need for stormwater runoff management, capture and transportation systems has developed due to human experiences with various challenges from destructive floods.

The target of sustainable stormwater management is to understand the changes in the urban landscape, in which vegetation addition is not widely observed, with the aim of devising approaches to limit certain undesirable effects and to take advantage of new opportunities (Huang et al 2007). A sustainable stormwater system is not a system to address runoff problems and avoid unwanted contaminants in the water, but rather, it is a system to increase the potential usability of water resources (Sundberg et al. 2004). Stormwater capture and drainage may be considered both systems to divert undesired water from urban areas and valuable elements for landscaping the surroundings of buildings and roads (Boller 2004). In general, flood control agencies have constructed large centralized facilities to control surface runoff, such as culverts, detention basins and sometimes re-engineered natural hydrologic features, including the paving of city river channels to quickly convey runoff to receiving water bodies. These large-scale facilities are required to handle the massive amounts of runoff generated by the largest storm events, as it would be impractical to handle such runoff on a decentralized parcel-by-parcel basis with small-scale infiltration devices. However, the current trend is toward a more integrated approach to manage stormwater runoff as an integrated system of prevention and control practices to accomplish stormwater management goals. The first principle is to minimize the generation of runoff and pollutants through a variety of techniques, and the second principle is to manage runoff and its pollutants to minimize their impacts on humans and the environment in a cost-effective manner (EPA 2007).

The utilization of remote sensing and GIS technologies in stormwater management is constantly evolving, and common GIS technologies are utilized to help decision-makers determine the most efficient ways to manage stormwater, including the selection of capture measures based on criteria and the evaluation of methods to capture urban runoff (Wilson et al. 2000). However, as explained by Yang et al. (2011), even in the most technically complex analyses, it is always necessary for the human element to select appropriate criteria and make other subjective decisions (Yang et al. 2011). Thus, decisions will frequently be made in stormwater management that reflect the economic, political, social and aesthetic components that may not always be easily incorporated into GIS analyses and modeling systems. Important aspects of community attitudes are also considered in this chapter, as these attitudes are influenced by a range of factors, including knowledge of urban water problems, their frequency and water restrictions, familiarity with the use of alternative water sources, and either positive or negative support from water authorities, government agencies and researchers (Dolnicar and Hurlimann 2009). Stormwater management in many countries like Japan and Thailand has always aimed to control stream flow for municipal and commercial use while preventing water-related disasters in cities. City stormwater management policies were historically based on flood control, but in recent years, they shifted from the exclusive use of structural approaches to using a combination of structural and nonstructural approaches/nature-based solutions. For instance, Japanese flood management began with policies implemented in 300 BC in the Yayoi Period, but in 1960, there was an effort to move away from concrete dams and instead focus on the hydrologic function of green dams, which rely on the flow-retarding capacity of forests to reduce flood risk (Luo et al. 2015; Calder 2007). On the other hand, Thailand has focused more on structural measures in addressing storm water issues, including the aggressive implementation of measures involving the general public and the private sector (Chiplunkar et al. 2012). In Vietnam, an integrated design for surface runoff infiltration, storage and transportation of stormwater has been applied, and the country has focused on increasing groundwater levels by increasing the stormwater percolation rates through enhanced infiltration (Werner et al. 2011).

This chapter presents the latest developments in stormwater management practices as a sustainable solution to achieve urban water security mainly in Asian region. For this purpose, the different characteristics of stormwater management policies and strategies adopted by Japan, Vietnam and Thailand are explored, and case studies from Tokyo, Hanoi and Bangkok are employed to evaluate the advantages and disadvantages of stormwater runoff practices and policies with respect to historical measures and the hydrologic dimensions of city stormwater management systems. This chapter provides a detailed assessment to assist future policy makers and researchers in understanding the significance and emerging role of stormwater management as a sustainable solution under the threat of future extreme events and climate change.

6.2.1 Urbanization

In the changing context of urbanization, climate change and various shifting management policies, stormwater management has been severely impacted. It is well known that urbanization alters watershed hydrology. As land becomes increasingly covered with surfaces impervious to rain, water is redirected from groundwater recharge, and climate change increases the evapotranspiration rate of stormwater runoff. As the area of impervious cover increases, so do the runoff volume and rate, and weak regulatory policies can worsen the situation (Schueler 1994; Corbett et al. 1997). Pollutants accumulate on impervious surfaces, and the increased runoff with urbanization is a leading cause of nonpoint source pollution (USEPA 2002). Sediment, chemicals, bacteria, viruses, and other pollutants are carried into receiving water bodies, resulting in degraded water quality (Holland et al. 2004; Sanger et al. 2008). Blair et al. (2011) tested and analyzed more than thirteen watershed locations in coastal Carolina (USA) and developed a method to model the impacts of urbanization and climate change on stormwater runoff that was based on the modeling methods of the Natural Resources Conservation Service (NRCS) and the United States Department of Agriculture (USDA), and they generated hydrographs of rate and time. The hydrograph in Fig. 6.2 shows the impact of different levels of urbanization on the rainfall percentage as well as the impacts of climate change on urban, suburban and forested watersheds.

In Fig. 6.2, the *y*-axis shows the runoff rate, and the *x*-axis shows time in hours. This hydrograph illustrates the impact of urbanization and climate change on runoff. Climate impact curves are based on a 5-inch rain in 24 h under semi-saturated runoff conditions (Blair et al. 2011). Considering changes in urbanization, climate and management policies, stormwater discharge pollution is versatile and arises from

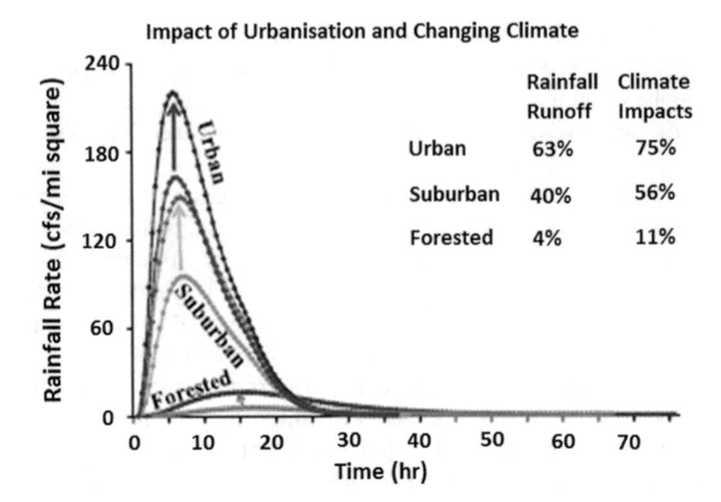

Fig. 6.2 Hydrographs showing runoff from three areas with different levels of development based on impervious cover. (*Source* Blair et al. 2011)

stormwater volume quantity and quality, and all the abovementioned factors that affect stormwater, which is derived from precipitation, such as rain, sleet or melting snow. In a natural setting, only a very small percentage of precipitation becomes surface runoff, but as urbanization increases, the stormwater percentage drastically increases. This runoff normally flows into the nearest stream or river, increasing the percentage of water in the system, but if it is polluted, it can lead to disastrous conditions and various forms of pollution for the receiving water bodies.

In the context of urbanization, we can accurately define stormwater as the runoff from pervious and impervious surfaces in predominantly urban environments. Impervious surfaces can be defined as concrete/charcoal roads, highways, roofs, pavement and footpaths. The land cover and precipitation relationship pathways have created a state in which watersheds and their streams and channels are adversely impacted (Frazer 2005), and in Fig. 6.1, we can see the relationship between watershed protection and urbanized areas covered with impervious roads and streams. Figure 6.2 demonstrates that increasing impervious surfaces alters the hydrologic cycle and creates conditions that can no longer support the diversity of life; the Center for Watershed Protection has reported that in areas that exceed 10% impervious coverage, stream health begins to decline (Coffman and France 2002). The problems faced by urbanized watersheds include flooding, stream bank erosion and pollutant export. The hydraulic characteristics of the recipient streams of these intensified storm flows are altered due to peak discharges several times higher than predevelopment or even rural land cover characteristics (LeRoy et al. 2006). Thus, improving the water quality of runoff entering receiving waters and reducing the pressure on existing water supply systems are major goals of urban water management.

6.2.2 *Climate Change*

In the context of climate change, which is considered to act in concert with urbanization, stormwater runoff and its impacts will likely intensify, further increasing the quantity of polluted runoff. Numerous studies involving climate change predictions have indicated that heavy precipitation events will likely increase in frequency and intensity (Karl et al. 2008). Semadeni-Davies et al. (2008) showed that the correlation between the increased intensity of rainfall and increased cover of impervious surfaces is directly proportional to more extreme events, such as flooding, flash floods and greater peak flows (Semadeni-Davies et al. 2008). Within the context of climate change, a science-based system for evaluating the relative impacts of both urbanization and climate change on stormwater runoff at a local scale is warranted. Stormwater management is planned based on local weather and climate, but climate changes, such as the amount, timing and intensity of rain events in combination with land development, can significantly affect the amount of stormwater runoff that must be managed. In some regions of the country, the combination of climate and land-use change may worsen existing stormwater-related flooding, whereas other areas may be minimally affected. In the past, stormwater management was practiced in an anthropocentric manner, which has had a profound effect on the environment. Figure 6.2 illustrates the correlation between climate change and rainfall, which is directly related to stormwater runoff.

Along with climate change, explosive suburban expansion over the last several years has increased the impervious surface cover in place of forests, pastures and cropland, which has affected local hydrological cycles by producing more surface runoff and decreasing the base flow, interflow and depression storage (Davis et al. 2006). Figure 6.3a shows the shift in climate and increase of global average temperature resulting in more hot extreme weather in future. Figure 6.3b is a time-series graph that compares the multi-model mean of the simulated global surface air temperature and precipitation from January 1979 to November 2012. The models suggest that manmade greenhouse gases have caused global surface air temperatures to warm and global precipitation to increase. Studies have shown a direct correlation between stream water quality and impervious surface coverage; with more than 10% impervious surface coverage, streams in some watersheds become unstable due to increasing damage from erosion or sedimentation (Vargas 2009).

As shown in Fig. 6.4, climate change and urbanization alter the physical factors associated with stormwater runoff and the responses of recipient waters.

6.3 Regulatory Policies

Sustainable stormwater management not only depends on the financial capacity of a country but on its policy priorities and institutional framework. Due to weak regulatory practices, many parts of the world are unable to sustainably manage stormwater,

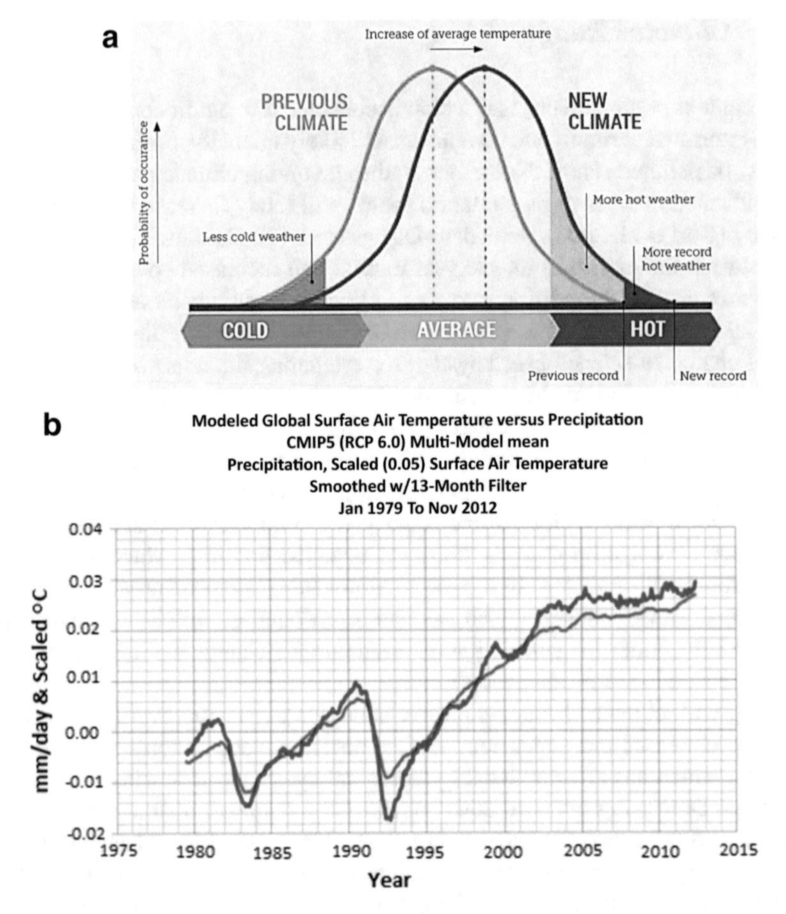

Fig. 6.3 **a** Shift in climate and the increase in the global average temperature (*Source* Climate commission 2013, redrawn from IPCC 2007). **b** Modeled global surface temperature with respect to precipitation (*Source* https://bobtisdale.wordpress.com/2012/12/27/model-data-precipitation-comparison-cmip5-ipcc-ar5-model-simulations-versus-satellite-era-observations/)

and studies have shown that stormwater runoff impacts both water quantity and quality. Excessive stormwater may result in stream bank erosion, thereby altering the stream bed morphology (Wynn 2004). During development, soils may become compacted, resulting in reduced infiltration capacity, which means that the amount of stormwater could possibly increase in quantity during construction because less rainwater is evaporating and infiltrating; this may lead to erosion, sedimentation, flooding, dissolved oxygen depletion, nutrient enrichment, reduced biodiversity, toxicity and other associated impacts on water use (Wagner et al. 2007). Water quality concerns vary from region to region depending on factors such as population density, land-use degradation and pollution (Fletcher and Deletic 2007). The major pollutants recognized in conventional stormwater management practices are suspended solids;

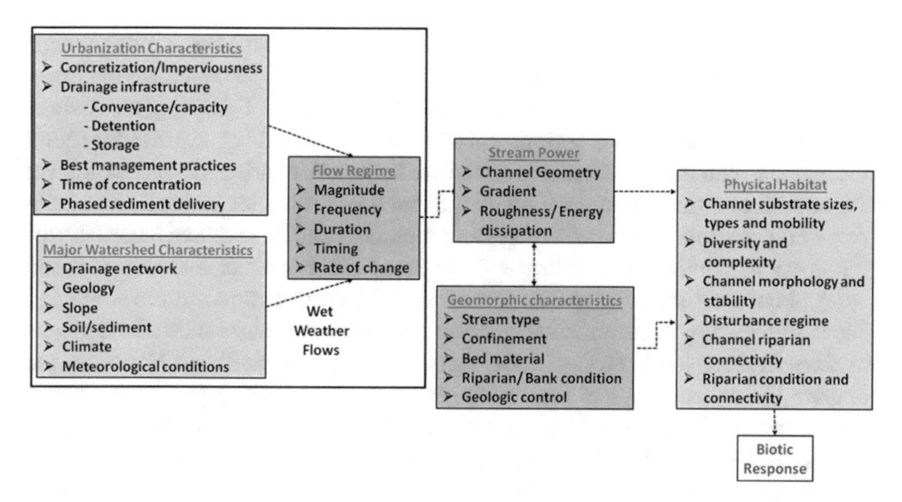

Fig. 6.4 Physical effects of urbanization on streams and habitats (*Source* Roesner and Bledsoe 2003)

oxygen-demanding matter; bacteria; nutrients, such as nitrogen and phosphorus; and heavy metals. Suspended solids increase turbidity, which directly affects ecosystems by lowering the amount of dissolved oxygen (DO) in the water (Bilotta and Brazier 2008); maintaining high DO levels in water is critical for sustaining aquatic life. The organic matter from animal feces and sewer overflows typically lowers DO levels in the recipient surface waters, and pathogenic bacteria in such overflow have caused detrimental human health effects as well as beach closures because of health safety issues, including many diseases associated with waterborne infections such as gastroenteritis and hepatitis (EPA 2001). Nitrogen and phosphorous mainly originate from agricultural fields, but they can also originate from the use of fertilizers and pesticides in an area (Adams and Papa 2000). Another form of pollution is thermal enrichment, which occurs when the surface water is heated and discharged into receiving water bodies with a lower temperature. This phenomenon can destructively affect water bodies with aquatic life that is sensitive to higher temperatures and reduce the dissolved oxygen concentration, which impacts the entire aquatic ecosystem.

6.4 Tools and Approaches in Optimal Stormwater Management

This section provides an overview of the environmental impacts of surface water runoff and the need for strong surface runoff management measures to achieve the sustainable use of water resources in urban environments. Previous studies have analyzed the impacts of stormwater runoff in changing environments, i.e., by comparing and evaluating findings under the contexts of climate change, urbanization and the stormwater regulation. Using hydrographs, it is easy to visualize the major negative impacts of surface runoff of nearby water bodies. Following an analysis of stormwater runoff under various contexts, an assessment of stormwater runoff management practices is discussed based on historical trends and considering the role of remote sensing and geographic information system technologies in identifying and determining stormwater capture measures and their effectiveness. Based on numerical simulation modeling and the economic aspects of stormwater runoff management measures, we assess current scenarios and optimal future strategies. This is followed by case studies to analyze current stormwater runoff management practices in three large cities in Asia, namely Tokyo, Bangkok and Hanoi, and to compare the advantages and disadvantages of policies with respect to historical, engineering and hydrologic dimensions. This chapter also presents a discussion of various best management practices (BMPs) and low-impact development (LID) practices. Finally, recommendations are suggested for policy-makers to design optimal stormwater runoff measures.

6.4.1 Historical Trends

In the ancient history of water management, which began thousands of years ago, the quality of life was directly correlated with flood control measures because floods had severe consequences in terms of human lives and crop production. The practices, techniques and strategies of ancient societies to control and manage stormwater can be studied (Koutsoyiannis et al. 2008), and approximately 3000 years ago, the ancient civilizations of Babylonia and Assyria were able to combine wastewater and stormwater sewage systems (Durrans 2003). In ancient times, water management served as a centralized issue because of the basic human needs for sanitation, clean drinking water and flood prevention, and these services have evolved into conventional stormwater management. To function efficiently, the conventional approach requires the construction of a massive and costly centralized infrastructure system, and many ancient stormwater drainage systems, such as the *Cloaca maxima* in Rome that was built in approximately 600 BCE, still exist (Fardin et al. 2014). Underground stormwater drainage was common in Europe and many places in the Americas in the nineteenth century following rapid urbanization (Burian and Edwards 2012) due to the Industrial Revolution. In particular, combined sewers can be observed in many cities, and the water carried in these systems has historically been directly transported into receiving water bodies. In ancient times, these approaches helped prevent flood damage as well as pollution, which often resulted in financial and environmental benefits, but in the beginning of the 1920s, flood prevention and stormwater management improved in linear fashion. It was assumed that stormwater was wastewater that needed to be transported outside of cities, and it was never considered as a resource (Durrans 2003). With the help of gravity, stormwater was easily disposed through sewerages to nearby water bodies, which inspired the development of urban drainage systems and the control of damage from infrequent flooding events. This was followed by designs that became common in every urban setting and that can be divided into two types of drainage systems: major systems designed to manage 100-year storm events and minor systems designed to manage 2–25-year storm events (Grigg 2012). Next came the development of detention ponds, a conventional yet inexpensive method to reduce peak flows and total volumes. However, detention ponds are limited in that they negatively impact the environment because they disrupt the drainage paths of streams and are unable to improve the quality of stormwater due to nonpoint pollution (Durrans 2003). In recent years, the development of new technologies and infiltration facilities changed the paradigm from ancient technologies with the one-dimensional view that stormwater is not useful and for which the goal management is only to reduce flood damage to an environmental insight that rainwater can be utilized as a valuable resource.

6.4.2 Potential Role of Remote Sensing and GIS Technologies in Stormwater Management

Remote sensing (RS) and geographic information system (GIS) technologies have been widely applied, and the integration of both is recognized as a powerful and effective tool for the design and formulation of stormwater management strategies. Remote sensing effectively collects multi-temporal, multi-spectral and multi-location data and facilitates observations of land-use changes, whereas GIS provides a platform for analyzing and displaying digital data and acts as a decision support system (Weng 2001). In urban environments, the first and most important management decision is identifying suitable stormwater harvesting sites for urban water management, and GIS has been recommended as a decision support tool (DSS) to facilitate this part of the decision-making process (Mbilinyi et al. 2005). GIS can also serve as a screening tool for the selection of preliminary sites as it has a unique capability for the analysis of multi-source data sets that allow for their integration (Malczewski 2004). Literature is widely available on the use of GIS to assess site suitability in terms of stormwater harvesting around the world. For example, potential sites for water harvesting in India were identified using the guidelines of the International Mission for Sustainable Development (IMSD) in a GIS environment (Kumar et al. 2008), and various studies have explained the development of GIS-based decision support systems (DSSs) to locate suitable sites for water harvesting (De Winnaar et al. 2007; Mbilinyi et al. 2007). Using a biophysical approach to understand the hydrological information derived from the characteristics of the catchment, criteria can be assessed to formulate a strategy to identify surface runoff harvesting sites (De Winnaar et al. 2007). Various studies have shown that areas with fewer spatial constraints for various water storages are suitable for stormwater harvesting based on such criteria as precipitation, runoff, soil type, topography and distance to storage (Mbilinyi et al. 2005; Kahinda et al. 2008). However, the spatial constraints are greater in an urban context due to the overall lower storage space and an already exhausted drainage network; this places constraints on the social, institutional and economic factors needed to locate suitable stormwater harvesting sites. For example, in Australia, Shipton and Somenahalli (2010) applied GIS to identify suitable stormwater runoff collection sites in the Central Business District of Adelaide, but the study was limited because it did not account for the demands of stormwater and only considered the drainage pattern in the region. The major finding was that GIS is a valuable tool to identify sites to build facilities related to stormwater management and the handling of surface runoff, and remote sensing provides the capability to design from a broader perspective rather than a narrow one. For effective stormwater runoff management, the use of specialized computer-based models in concert with GIS can produce better results (Rusko et al. 2010) explained that the integration of simulations with sustainability objectives involving social, environmental and socioeconomic factors into storm water management policies could prove useful. For effective storm water management at the local level, this approach would be useful along with GIS and remote sensing. It is easier for policy-makers to implement policies using different areas in the watershed rather than different, locally adaptable strategies to solve locally variable problems such as erosion, sedimentation, flooding or pollution, and in this respect, GIS and remote sensing are highly applicable. This allows for a complete stormwater management plan across the watershed to prevent the negative effects of stormwater at specific sites and anywhere downstream where there is potential for harm, which can be easily identified by monitoring the sites using a remote-sensing map and processing the data through GIS. Another use of GIS and remote sensing is to collect and manage spatial data, which is an important requirement as such data are used as inputs for computer stormwater models such as MOUSE, MIKE II, Hydroworks, SWMM and STORM (Elliott and Trowsdale 2007). GIS can then be used to present the user-friendly processed result of the model outputs (Heaney et al. 2001). Modeling using GIS to handle data from computer-based tools to make a user-friendly decision support system can improve communication among stakeholders involved in modern-world stormwater planning. GIS and remote sensing for stormwater runoff management are used for flood mapping, flood hazard mapping, hydrologic modeling, catchment-level stormwater management and the design of storm sewer systems, including the determination of the slope and surface elevation from digital elevation model (DEM) data. Moreover, they can be used for stormwater management planning, such as identifying best management practices, low-impact development and feasibility assessment (Rusko et al. 2010).

The use of GIS and remote sensing to estimate stormwater runoff from the land use, slope, impervious surface coverage and soil characteristics is common and evolving, and it can aid decision-makers in deciding the optimal way to manage stormwater planning decisions. However, it is important to note that for any complex situation or analysis, it is still necessary to have a human element to select the appropriate criteria and make subjective decisions; frequent decisions are required in stormwater management. Another widely applicable use of remote sensing is the control of nonpoint source pollution by analyzing the impervious surfaces within watersheds (Slonecker et al. 2001), and it is also useful for estimating stormwater pollutant mass loading (Ackerman and Schiff 2003).

6.4.3 Numerical Simulation Modeling

With rapid urbanization and climate change, stormwater managers face increasingly complex issues regarding the design, construction, operation and maintenance of different stormwater infrastructures. Recent stormwater management systems have largely focused on ecosystem-based approaches (instead of traditional approaches of moving the stormwater out of the affected area), which require groundwater recharge, maintenance of the natural flow regime, and consideration of the downstream impacts and water quality. With the advent of computer systems and tools such as GIS and remote sensing, numerical models have become more widely used to simulate hydrologic–hydraulic processes by mathematically representing the stormwater movement systems. Numerical models enable the effectiveness of different alternative stormwater management measures to be tested by simulating water quantity and quality values at different locations, and they are principally based on a set of differential equations representing the physical stormwater management systems. Furthermore, they describe the rates of change of various parameters with respect to time and space, so running a numerical model implies solving these equations with boundary conditions and spatial/temporal changes within the systems. The model results represent the responses to global changes (e.g., climatic, land use and population growth) and alternative management measures. Examples include graphs of depth or flow at specific locations in the network.

Figure 6.5 is a collection of urban stormwater models used for stormwater simulation and management. Figure 6.5a shows the list of potential uses of the listed urban stormwater models; e.g., the model MOUSE is applicable in developing sizing rules for devices, planning and land use in catchments and detailing in the design of site layouts. Figure 6.5b shows the various analyzed urban stormwater models based on their spatial and temporal resolutions. Finally, Fig. 6.5c shows the capability of the model in simulating runoff generation with different routing methods.

6.4.4 Economic Assessment

Economic assessment of stormwater runoff management is important in improving environmental quality; the basic concept is to have an incentive-based market to understand stormwater as a resource rather than waste. In this section, cost-effective measures to control stormwater runoff by providing incentives for small-scale best management practices (BMPs) throughout urban watersheds are explored (Parikh et al. 2005), and the solution to reducing stormwater runoff and other impacts is a market mechanism within the incentive-based market, such as stormwater runoff user fees, charges, cap and trade, and voluntary offset programs (Parikh et al. 2005). Many other solutions have been proposed, one of which is to pay for pollution, which is a basic strategy that can be achieved by measuring the pollutants in an environment and charging the culprits at the source (Dales 2002); another strategy is the

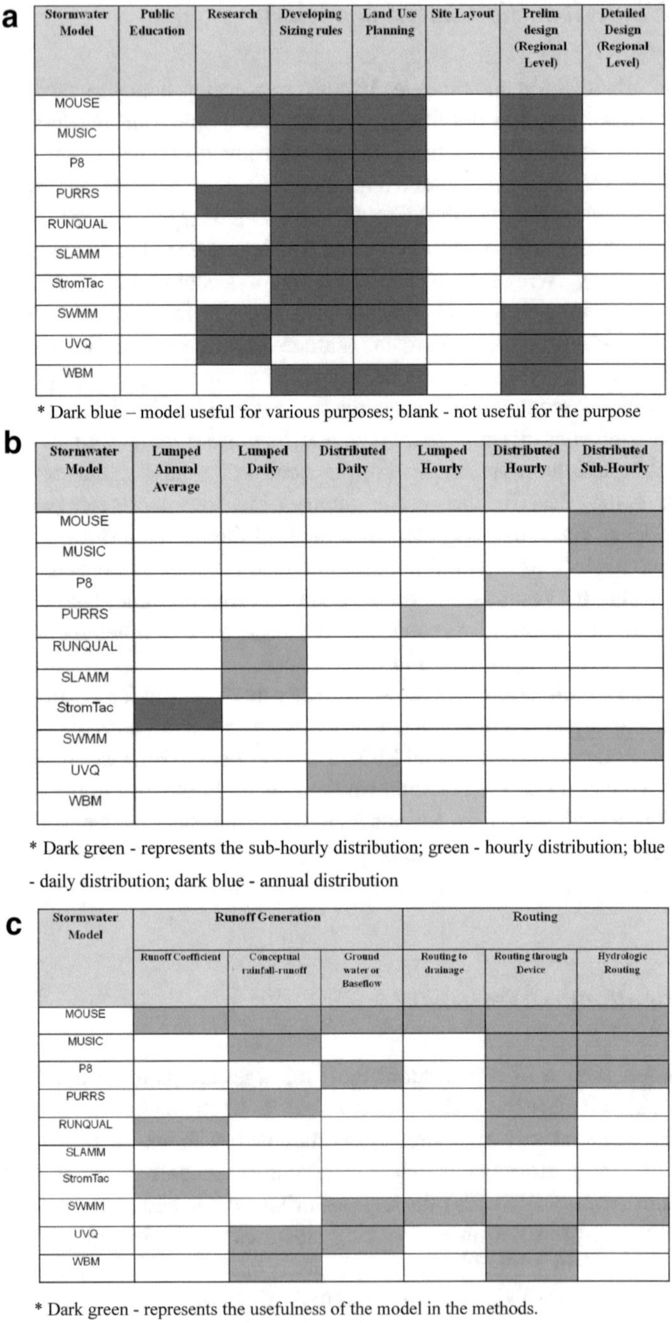

a

Stormwater Model	Public Education	Research	Developing Sizing rules	Land Use Planning	Site Layout	Prelim design (Regional Level)	Detailed Design (Regional Level)
MOUSE							
MUSIC							
P8							
PURRS							
RUNQUAL							
SLAMM							
StromTac							
SWMM							
UVQ							
WBM							

* Dark blue – model useful for various purposes; blank - not useful for the purpose

b

Stormwater Model	Lumped Annual Average	Lumped Daily	Distributed Daily	Lumped Hourly	Distributed Hourly	Distributed Sub-Hourly
MOUSE						
MUSIC						
P8						
PURRS						
RUNQUAL						
SLAMM						
StromTac						
SWMM						
UVQ						
WBM						

* Dark green - represents the sub-hourly distribution; green - hourly distribution; blue - daily distribution; dark blue - annual distribution

c

Stormwater Model	Runoff Generation			Routing		
	Runoff Coefficient	Conceptual rainfall-runoff	Ground water or Baseflow	Routing to drainage	Routing through Device	Hydrologic Routing
MOUSE						
MUSIC						
P8						
PURRS						
RUNQUAL						
SLAMM						
StromTac						
SWMM						
UVQ						
WBM						

* Dark green - represents the usefulness of the model in the methods.

Fig. 6.5 **a** Potential uses of selected models. Gray shading indicates that the model is marginally suited to that use (*Source* Elliott and Trowsdale 2007). **b** Temporal and spatial resolutions of the urban stormwater models (*Source* Elliott and Trowsdale 2007). **c** Runoff generation and routing methods for the urban stormwater model (*Source* Elliott and Trowsdale 2007)

promotion of technological advancements to increase stormwater conservation and prevent pollution. Stormwater controls can be created using a fee structure that is directly related to the degree of water degradation, i.e., 'pay for your pollution'; this forces polluters to provide compensation for damage to the environment in any form with the goal of preventing pollution as the first priority. However, some studies have shown that existing fees do not have the necessary impact to lead to on-site reductions, and in a few cases, they are not sufficient to stop polluters from polluting (Doll et al. 1999). These studies indicated that the fees were too low to have an impact on stormwater runoff control. Another method is to utilize stormwater to generate revenue to combat stormwater degradation in relation to the impervious surface coverage at any level (Cyre 2000); generally, nonpoint source constituents are not regulated by any mechanism because they are unobservable in nature. Therefore, a popular method for stormwater runoff control is a stormwater trading mechanism (Thurston 2006), which some communities have implemented by calculating a flat rate based on imperviousness, and the money that is generated is used to develop infrastructure for stormwater control, which requires the involvement of local and regional partners to invest in groundwater infiltration facilities and stormwater technologies, such as LID and BMPs (Doll et al. 1999). Trading mechanisms provide credit-trading incentives to manage sustainable stormwater use and allow developers and designers to protect the overall quality of water bodies (Woodward and Kaiser 2002). In addition to water temperature and nutrient loading, the total stormwater produced in an urban area and its sustainability values (i.e., economic, environmental and social) are important considerations for trading schemes. By using a market-based strategy to control stormwater and improve water quality, large polluters can purchase water quality improvements from smaller producers as a voluntary runoff offset to optimize economic metrics.

To control and manage stormwater runoff, BMPs are considered less expensive than conventional centralized systems and more efficient in improving water quality, but they require considerable investment. Analyzing their cost effectiveness is a complicated process. Cost-effectiveness analysis identifies the least expensive way of achieving environmental targets (Ecosystem Valuation 2007), and the five techniques used to measure the value of BMPs are replacement cost methods, life-cycle cost analysis, cost-benefit analysis, the productivity method and the hedonic pricing method. In the replacement cost method, the valuation of ecosystem services is performed based on the cost of avoiding damage or lost services. For example, in 2006, researchers at the University of California, Davis estimated that for every 1000 trees in the central valley, stormwater runoff was reduced by nearly 1 million gallons, equivalent to 7000 USD. When those trees are cut down and their function is lost due to deforestation, the costs are passed to the local government (Kloss 2006). The life-cycle approach is useful in estimating the BMP costs, because most of the time, the maintenance costs are far less than the cost involved in maintaining and operating conventional approaches (Powell et al. 2005). In cost-benefit analysis, the lack of information about the economic benefits provided by the BMPs can ultimately obstruct their adoption and implementation (MacMullen 2007). Productivity valuation, which is also known as the net factor income, can be used to estimate the economic value of the ecosystem products that contribute to the production of marketed goods. Lastly, the hedonic pricing method estimates the economic values for environmental services that directly affect the market price, and it is widely used to determine variations in housing prices while reflecting the values of local environmental attributes (Ecosystem Valuation 2007).

6.5 Urban Stormwater Governance: Case Studies

Cities in Japan, Vietnam and Thailand are eagerly applying stormwater infiltration as a stormwater runoff management measure, but many of the local governments continue to hesitate to ask residents to install soak ways and small detention basins on private properties. Based on the above discussion of stormwater runoff control management, the case studies of three megacities in Asia are discussed. It appears that stormwater infiltration is widely accepted and implemented all over Japan, but in other countries, such as Thailand and Vietnam, it remains unpopular. Along with developing stormwater infiltration measures, governments have focused on building massive sewerages to capture and transport stormwater out of the city in case of flooding or emergencies. For our case studies, we chose Tokyo, Bangkok and Hanoi to analyze and compare stormwater runoff management practices to understand the optimal strategies and thus provide recommendations.

6.5.1 Tokyo

Tokyo is the capital city of Japan with a total area of 2188 km^2, and it is situated at 35° N latitude and 139° E longitude. The annual precipitation in Tokyo is approximately 1530 mm, and it has the largest population of any city in the world. The city also suffers from serious flood problems, and stormwater runoff has always been serious concern for the local government. As shown in the land-use and land cover map in Fig. 6.6, Tokyo has undergone substantial urbanization, and the city is focused on structural and nonstructural measures for stormwater runoff control. For example, Tokyo has been investing in and building huge tunnels, such as G-Cans, and the local government has been simultaneously encouraging residents to use BMPs and groundwater infiltration systems.

To address flooding concerns and the exacerbating effect of stormwater runoff and torrential rains, Tokyo has invested in underground infrastructure, using five silos and through-tunnel channels to transport water out of the city. This is referred to as the Metropolitan Area Outer Underground Discharge Channel or G-Cans project, and it is the largest underground flood water diversion facility in the world (Bobylev 2007) (Fig. 6.7a and b). It is located in the outskirts of the city between Showa in Tokyo and Kasukabe in Saitama Prefecture and is an example of a critical underground infrastructure for a flood and stormwater control management system. The figure

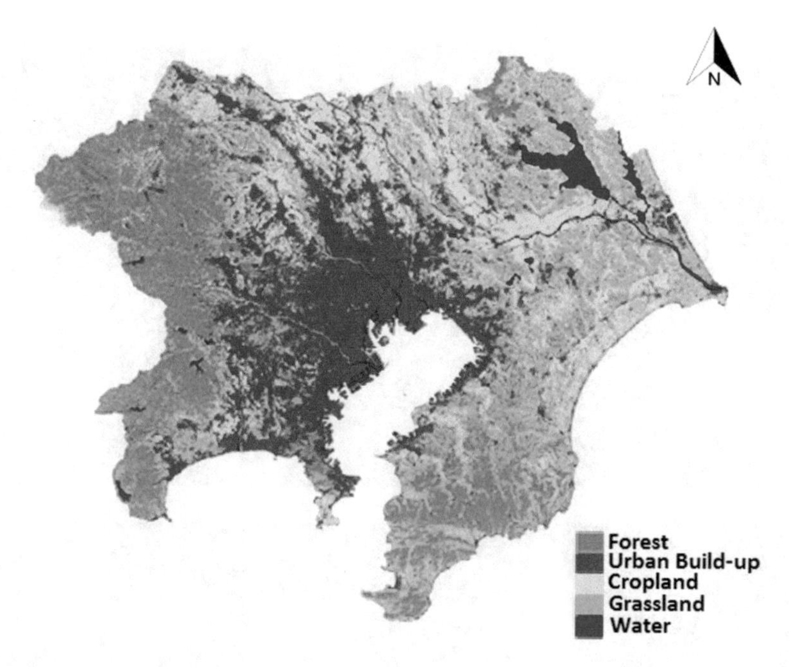

Fig. 6.6 Tokyo land-use and land cover maps

shows the wide network of tunnels used to manage the stormwater by draining surface runoff out of the city.

The project includes a 6.5-km-long connecting tunnel and a storage tank with 78 pumps and five massive silos. There are five concrete containment silos with dimensions of 65 m in depth by 32 m in diameter that act as a regulator, and they are located at the limits of the rivers to connect to the tunnel. The tunnel was constructed underground at a depth of approximately 50 m to send the water to the storage tank during overflow conditions. The storage tank, also known as the underground temple, is 177 m long and 25 m high and supported by 59 pillars that are each 20 m tall. The tank has approximately 14000 turbines. This facility is also open to tourists during the dry season (G-Cans project, Tokyo).

Another very popular and highly focused stormwater control measure is the artificial infiltration stormwater system in the city. Figure 6.8 shows the structure, in which permeable pavement leads to a facility with an artificial infiltration trench that is connected to the combined sewer in the case of over flow. In addition of flood prevention, this provides clean water to the environment and a functional water cycle with a recharging groundwater table that can restore the spring water and allow for disaster prevention and preservation of local ecosystems. Overall, the stormwater runoff control management practices in Tokyo are highly effective.

Fig. 6.7 **a** and **b** G-Cans in Tokyo is an underground infrastructure to prevent flooding prevention during the rainy season (*Source* G-Cans project, Tokyo; http://www.g-cans.jp/)

6.5.2 Bangkok

Bangkok is the capital of Thailand. The total area of the city is 1569 km^2, and it is located at 13.45° N latitude and 100.28° E longitude. The mean elevation is 2.31 m above the mean sea-level (BMA and UNEP 2004). In total, 60% of the land area of the city is built up, and approximately 30% is utilized for agriculture. In 2015, the total population of the city was approximately 8,500,000 with a growth rate of

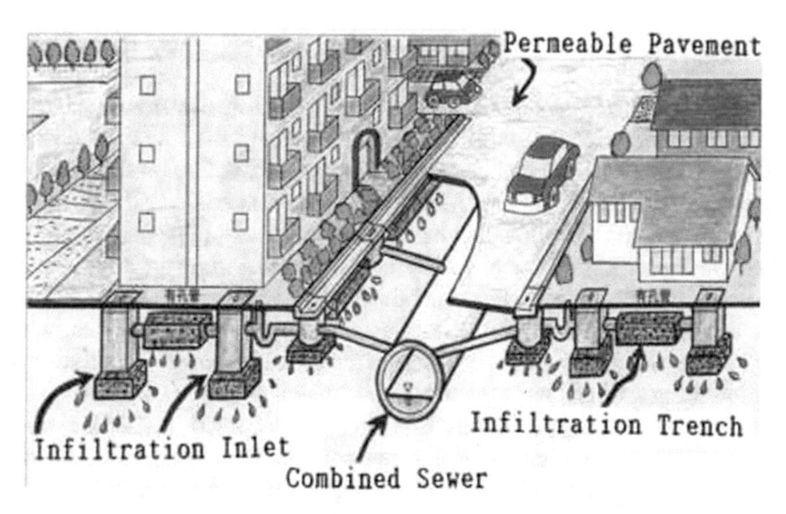

Fig. 6.8 Infiltration trench and sewerage system in Tokyo (*Source* http://www.recwet.t.u-tokyo.ac.jp/furumailab/crest/workshop05/june9pm_2.pdf)

less than 2% annually (United Nations, Department of Economic and Social Affairs, Population Division, 2014) (Fig. 6.9). The important seasons in Bangkok include the rainy season from May to October, winter from November to January and summer

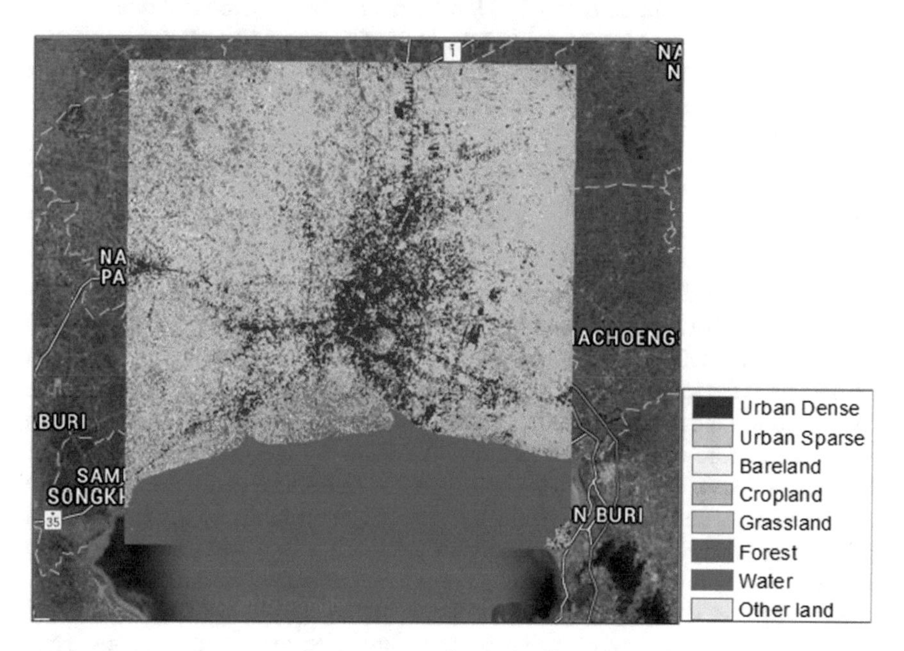

Fig. 6.9 Land-use and land cover maps of Bangkok in 2014 (*Source* http://giswin.geo.tsukuba.ac.jp/capital-cities/)

from February to April. The average annual precipitation is approximately 1500 mm (Shrestha et al. 2015).

Bangkok is facing severe challenges in terms of stormwater runoff management that are mostly interrelated with the impact of urbanization and the changes in hydrology associated with global climate change (IPCC 2007). To address such stormwater issues, the Bangkok Metropolitan Administration (BMA) has been focused on structural measures and has been aggressively implementing measures involving the general Bangkok populace that emphasize strong drainage provisions, such as retention ponds and large-scale rainwater harvesting, to alleviate stormwater runoff, especially in public and private establishments. Another focus, which is slowly building in momentum, is recharging the groundwater bodies via stormwater infiltration using artificial recharge techniques (Chiplunkar et al. 2012).

In addition to structural measures, best management practices (BMPs) which are also focusing on flood mitigation of stormwater runoff help combat flash flooding in urban areas and groundwater infiltration. Among various BMPs, vegetative swale, bio retention, vegetative strips along roads, ponds and permeable pavements are quite popular. Critical underground infrastructure is gaining popularity in Bangkok as previous flood effects were disastrous, and the city is currently building comprehensive surface runoff and flood control systems already. Additionally, on a governance level, the use of multipurpose critical underground drainage facilities for stormwater management has been gaining support (Broere 2016). The plan is to build a double deck cut and cover the infrastructure beneath the existing eight-lane eastern outer ring road, which is approximately 100 km from a northern suburb and runs parallel to the river to drain the extra water into the Gulf of Thailand (Article: Tunnel Talk). In Fig. 6.10, we can see a schematic diagram of the plan to build the underground infrastructure.

6.5.3 Hanoi

Hanoi is the capital city of Vietnam and is surrounded by the Red River, the Nhue River and the To-Lich River, and it has an area of approximately 925 km^2 (Duan and Shibayama 2009). The population of the city is estimated to exceed 4 million inhabitants as the four main precincts have each exceeded 1 million. According to the Hanoi Master Plan for the year 2020, calculations show an increase in the population of the urban core to nearly 4.5 million inhabitants by the year 2020 (Duan and Shibayama 2009). This forecast is more than just an estimate; with the current rate of urbanization, this is a very realistic scenario. Figure 6.11 shows the land use and land cover of Hanoi, and the effect of urbanization on the land use of the city is easily observed.

According to a study by the World Bank, the country ranked among the top five countries that are going to be severely impacted by climate change, especially in terms of sea-level rise and flooding (Dasgupta et al 2007). The common stormwater management practices in Hanoi are surface runoff drainage, rainwater

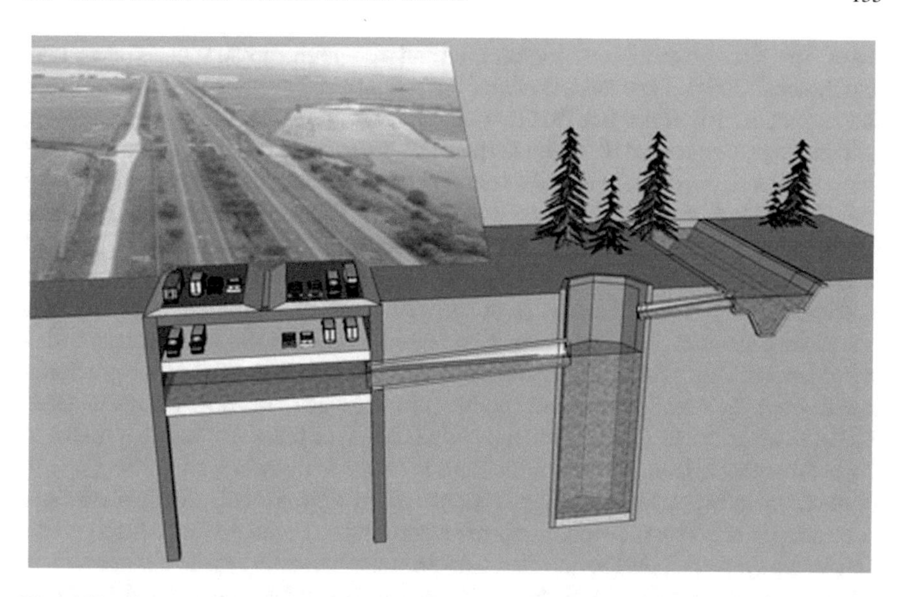

Fig. 6.10 Plan to build underground infrastructure to mitigate floods and stormwater surface runoff (Image adopted from http://www.tunneltalk.com/Bangkok-Thailand-Dec11-Floods-bring-forward-major-mitigation-plans.php)

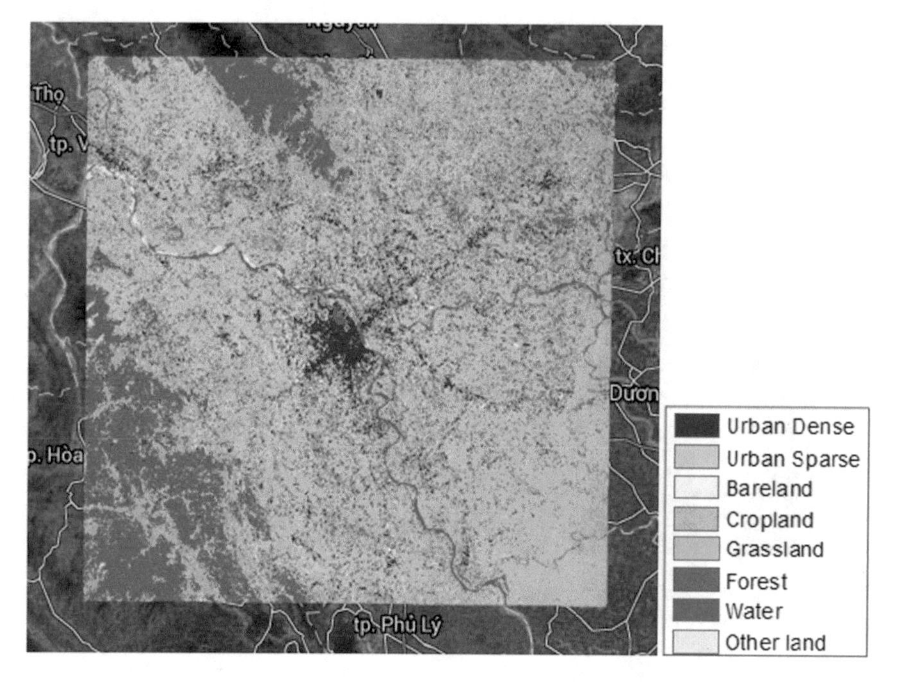

Fig. 6.11 Land-use map and land cover map of Hanoi in 2014 (*Source* http://giswin.geo.tsukuba.ac.jp/capital-cities/)

harvesting, subsurface rainwater infiltration and storage and urban aquaculture (Hoan and Edwards 2005). In general, decentralized stormwater management measures are incorporated into the design of the urban landscape to tackle the problem at the source, and drainage systems are designed to transport the runoff to the nearest stream. The urban aquaculture system in Hanoi is also very useful in addressing excess stormwater runoff (UNEP, International Source Book on Environmentally Sound Technologies for Wastewater and Stormwater Management). As shown in Fig. 6.12, a good practice is to avoid stormwater runoff and mitigate its negative effects, but another approach is urban forestry within the city, as urban plantations can mitigate the stormwater runoff management problems caused by impervious surfaces. It is estimated that trees slow the flow of stormwater runoff by absorbing the first 30% of the precipitation through their leaf system and another 30% through their root systems (Burden 2006); in addition, trees help filter the pollutants out of the water before it drains to the nearby streams. An important measure in Hanoi is rainwater harvesting, as can be observed in Fig. 6.13; the design is adopted from Nguyen et al. (2013). Rainwater harvesting is effective in reducing stormwater runoff in Hanoi due to its high precipitation. Nguyen et al. (2013) installed ten rain water harvesting machines at low cost using a canvas catchment, a plastic tank and a stainless-steel tank; Fig. 6.13 shows the stainless-steel tank, which is very useful in reducing and controlling stormwater (Nguyen et al. 2013).

Fig. 6.12 Raw wastewater- and stormwater-fed fish ponds in Hanoi, Vietnam (*Source* http://www.unep.or.jp/ietc/Publications/TechPublications/TechPub-15/2-9/9-3-1.asp)

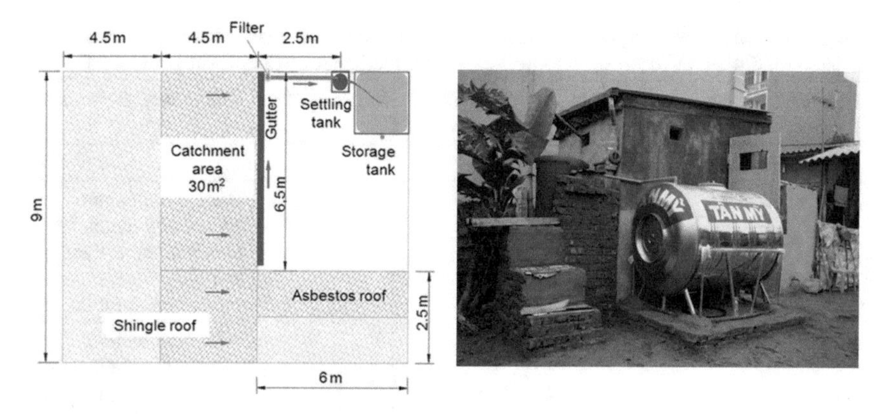

Fig. 6.13 Rainwater harvesting practices in Hanoi (*Source* Nguyen et al. 2013)

6.6 Summary

Stormwater runoff management practices can be broadly classified into two categories. The first is the reduction of surface water runoff quantity, and the second is the improvement of stormwater quality before it drains into nearby water bodies or infiltrates into the ground. Stormwater runoff management practices have evolved over time, and new strategies have been developed to minimize the negative impacts of stormwater runoff. These strategies are known as LID or low impact development, which involves a combination of sound site planning, structural techniques and nonstructural techniques (Dietz 2007). LID techniques promote nature-based designs and interact with the processes that control stormwater runoff, thereby reducing its negative impacts (Davis and Richard 2005). Typical elements of conveyance are pavement, roofs, pipes and lawns. Studies have revealed an increasing trend toward more sustainable practices, and it is clear that conventional practices are not sustainable and cause environmental degradation (Villarreal 2004). Therefore, a shift to LID or best management practices to mitigate the effects of stormwater is warranted. Effective LIDs include runoff mitigation measures that are part of larger group of strategies and practices in stormwater management known as best management practices or BMPs (Perez-Pedini et al. 2005), and the combination is known as LID-BMPs or low-impact development-based BMPs. Table 6.1 presents a list of BMPs to demonstrate the effectiveness of different strategies in terms of managing stormwater runoff.

Nonstructural BMPs include practices such as reducing the disturbance at a given site, preserving important features, reducing impervious cover, incorporating vegetation options and focusing on natural drainage. Nonstructural BMPs are generally classified into four categories including landscaping and vegetative swales, minimizing the disturbance at a given site, management of impervious areas and modification of stormwater concentrations. Alternatively, structural BMPs operate very close to the runoff sources and are used to control and treat runoff. Typical structural BMPs include filters and surface devices located on individual lots in residential,

Table 6.1 BMP Table—type and effectiveness of different best management practices (BMPs) including infiltration BMPs (*Source* http://www.coralreef.gov/transportation/evalbmp.pdf)

Best management practices	Volume	Peak discharge	Water quality	Unit operation/process
Low-impact development techniques				
Bioremediation	†	‡	‡	Volume reduction; microbially mediated transformation; uptake and storage; size separation; sorption
Bioslope	†	†	‡	Volume reduction; microbially mediated transformation; uptake and storage; size separation; sorption
Catch basin controls	**0**	**0**	•	Size separation and exclusion; density, gravity, inertial separation
Gutter filter	**0**	•	†	Size separation and exclusion; physical sorption
Infiltration trenches/strips	†	†	†	Volume reduction; size separation and exclusion; chemical sorption processes
Permeable pavement	‡	‡	†	Volume reduction; size separation and exclusion
Pollution prevention/street sweeping	**0**	**0**	• /†	N/A
Surface sand filter	•	•	†	Size separation and exclusion; microbially mediated transformation; sorption
Soil amendments	†	†	†	Volume reduction; size separation and exclusion; microbially mediated transformation; uptake and storage; sorption
Swales	†	†	‡	Volume reduction; density, gravity, inertial separation; microbially mediated transformation; sorption

(continued)

Table 6.1 (continued)

Best management practices	Volume	Peak discharge	Water quality	Unit operation/process
Vegetation/landscaping	‡	‡	‡	Volume reduction; microbially mediated transformation; uptake and storage
Conventional and innovative techniques				
Advanced biological systems	†	†	‡	Microbially mediated transformation; uptake and storage
Detention and retention ponds	**0**	‡	•	Flow and volume attenuation; density, gravity, inertial separation; coagulation/flocculation
Disinfection systems	0	0	†	Chemical disinfection
Flocculent/precipitant Injection	0	0	†	Coagulation/flocculation
Sedimentation ponds and forebays	0	†	•	Flow and volume attenuation; density, gravity, inertial separation
Surface filters (filter ffabrics)	0	0	•	Size separation and exclusion
Disturbed soil restoration	†	•	†	Size separation and exclusion; density, gravity, inertial separation; microbially mediated transformation
Rain garden	•	•	†	Flow and volume attenuation; density, gravity, inertial separation
Pocket wetland	0	‡	†	Flow and volume attenuation; density, gravity, inertial separation; microbially mediated transformation; coagulation/flocculation

0 No impact, • Less effective, † Moderately effective, ‡ Highly effective

commercial or industrial developments (Homer et al. 2004). Figure 6.14 shows all the types of structural BMPs, including cisterns, rain barrels and vertical storage devices, that are used to capture rainwater. Cisterns are used for holding larger volumes of water, i.e., approximately 500 gallons or more, and rain barrels normally collect drainage from roofs using pipe networks and are particularly useful in small irrigation units. Vertical storage involves standing towers against a building to capture

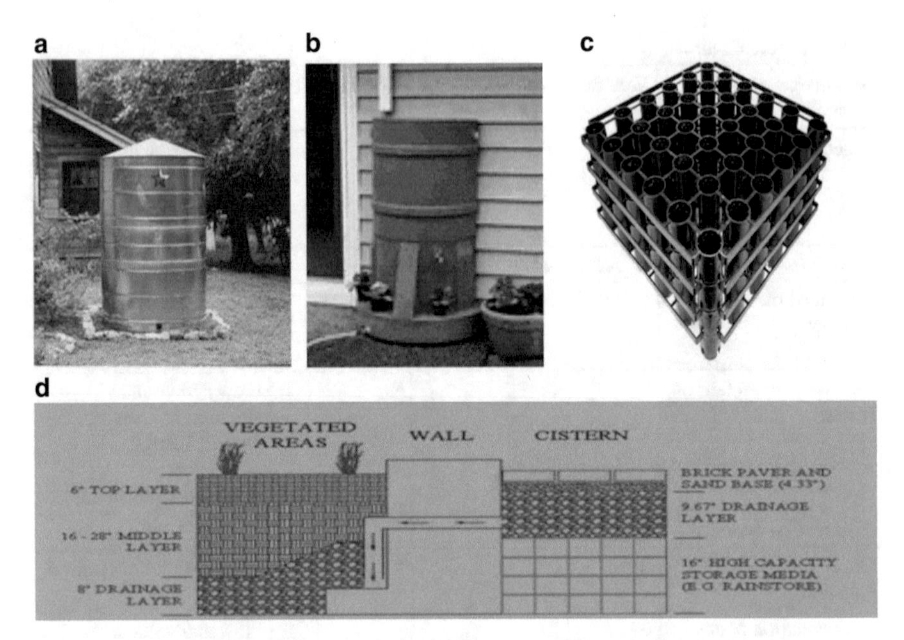

Fig. 6.14 Structural best management practices for stormwater runoff management: **a** cisterns, **b** rain barrels, **c** vertical storage structures and **d** storage beneath a structure (*Source* http://www.elibrary.dep.state.pa.us/dsweb/Get/Document-68001/6.5.2%20BMP%20Runoff%20Capture%20and%20Reuse.pdf)

and store water. Another interesting form involves storage beneath the structure, by which the stormwater is stored inside the structure (Guo and Baetz 2007). Another stormwater management approach is a drainage system to take the stormwater out of the city to the nearest water bodies; e.g., Japan built a massive drainage structure. The abovementioned LID and BMP techniques integrated with bioretention basins and structural BMPs are applied in stormwater management to improve the quality of stormwater. A new paradigm of infiltration systems, bioretention basins and surface and subsurface detention basins has developed to address stormwater runoff close to the source.

Green infrastructure is also an important part of management, and it is comprised of interconnected networks of natural areas, such as forests and wetlands, that help to improve water quality and provide ecosystem services in the form of recreational activities, wildlife habitat and good air quality. The important aspect of the discussion here is that community acceptance and governance of stormwater runoff management are equally important. As shown in Table 6.2, different entities become responsible for governance at various levels. At the international level, policies are formed based on agencies such as the UN and other NGOs, and on the national level, the governance structure and policies are formed by environmental agencies and legislative bodies. Proper maintenance of stormwater facilities can lead to reduced costs for stream channel restoration and pollution mitigation in the future. Ferguson (2005) described

Table 6.2 Structural governance level and the responsible entities for stormwater management (redrawn from Porse 2013)

	Governance level	Responsible entities
1.	International	UN Agencies Nongovernmental organizations
2.	National	Environmental agencies Flood management agencies Legislative bodies
3.	State/Territorial	Water management agencies Environmental agencies
4.	Regional	Regional government councils Issue-focused, multiagency entities

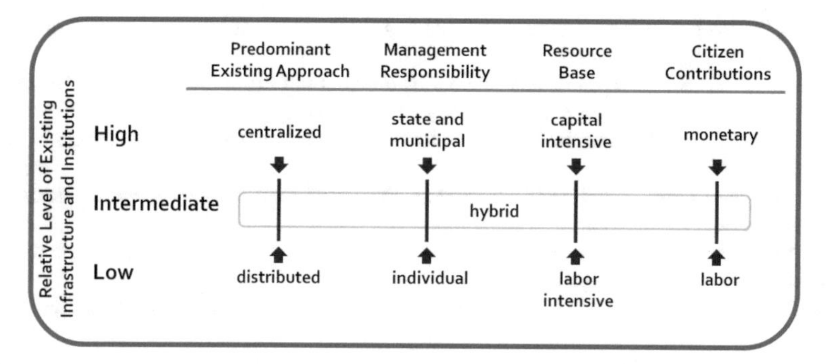

Fig. 6.15 Governance structure for stormwater management in cities (*Source* Porse 2013)

the conventional engineering methods used to mitigate the effects of stormwater by quoting the oldest method, i.e., reducing the nuisance of water by providing a smooth surface. Figure 6.15 shows how a hybrid governance structure manages stormwater; as infiltration-based approaches increase, cities have to share the responsibility by building and maintaining green roofs, rain gardens and swales on private properties. Local municipalities can oversee public communication, incentive programs and monitoring (Porse 2013). This system is considered better, in many respects, than federal regulations supporting a centralized system to mitigate stormwater runoff because in hybridization management, the citizens of the cities have the opportunity to determine the governance, economic and social aspects of stormwater management.

References

Ackerman D, Schiff K (2003) Modeling stormwater mass emissions to the Southern California Bight. J Environ Eng 129:308–317

Adams BJ, Papa F (2000) Urban stormwater management planning with analytical probabilistic models. Wiley, New York, USA

Ashley RM, Balmforth DJ, Saul AJ, Blanksby JR (2005) Flooding in the future. Water Sci Technol 52(5):265–274

BMA and United Nations Environment Programme (UNEP) (2004) Bangkok State of Enviornment 2003

BMA and UNEP Regional Office in the Asia and Pacific, pp 2004. www.rrcap.unep.org/pub/soe/bangkoksoe03.cfm

Bahri A (2012) Integrated urban water management. GWP technical background papers. Global water partnership, Stockholm Number 16. Climate change and water. Technical Paper of the Intergovernmental Panel on Climate Change. In: Bates BC, Kundzewicz ZW, Wu S, Palutikof JP (eds) IPCC Secretariat, Geneva 210 pp

Bilotta GS, Brazier RE (2008) Understanding the influence of suspended solids on water quality and aquatic biota. Water Res 42(12):2849–2861

Blair A, Sanger D, Holland AF, White D, Vandiver L, White S (2011) Stormwater runoff? Modeling impacts of urbanization and climate change, Louisville, Kentucky, 7–10 Aug 2011. Am Soc Agric Biol Eng

Broere W (2016) Urban underground space: solving the problems of today's cities. Tunn Undergr Space Technol, 55, pp 245–248

Bobylev N (2007) Sustainability and vulnerability analysis of critical underground infrastructure: managing critical infrastructure risks. Springer, Netherlands, pp 445–469

Boller M (2004) Towards sustainable urban stormwater management. Water Sci Technol. Water Supply 4(1):55–65

Burden D (2006) Urban street trees: 22 benefits and specific applications. Glatting Jackson and Walkable Communities Inc.

Burian SJ, Edwards FG (2012) Historical perspectives of urban drainage. In: 9th international conference on urban drainage, pp 1–16

Calder IR (2007) Forests and water—ensuring forest benefits outweigh water costs. For Ecol Manag 251:110–120

Chiplunkar A, Kallidaikurichi S, Cheon KT (2012) Good practices in urban water management: decoding good practices for a successful future. Asian Development Bank, Mandaluyong City, Philippines

Coffman LS, France RL (2002) Low-impact development: an alternative stormwater management technology. Handbook of Water Sensit Plan Des 97–123

Corbett CW, Wahl M, Porter DE, Edwards D, Moise C (1997) Nonpoint source runoff modeling a comparison of a forested watershed and an urban watershed on the South Carolina coast. J Exp Mar Biol Ecol 213:133–149

Cyre HJ (2000) The stormwater utility concept in the next decade. EPA National Conference on Tools for Urban Water Resource Management and Protection, U.S. EPA Office of Research and Development, USA

Dales JH (2002) Pollution, property & prices: an essay in policy-making and economics. Edward Elgar Publishing, pp 117

Dasgupta S, Laplante B, Meisner C, Wheeler D, Yan J (2007) The impact of sea level rise on developing countries: a comparative analysis. World Bank Policy Research Working Paper 4136

Davis AP, Richard HM (2005) Stormwater management for smart growth. Springer Science & Business Media, pp 368

Davis AP, Shokouhian M, Sharma H, Christie M (2006) Water quality improvement through bioretention media: nitrogen and phosphorus removal. Water Environ Res 284–293

De Winnaar G, Jewitt GPW, Horan M (2007) A GIS-based approach for identifying potential runoff harvesting sites in the Thukela River basin South Africa. Phys Chem Earth Parts A/B/C 32(15–18):1058–1067

Dietz ME (2007) Low impact development practices: a review of current research and recommendations for future directions. Water Air Soil Pollut 186(1–4):351–363

Doll A, Scodari PF, Lindsey G (1999) Credits as economic incentives for on-site stormwater management: issues and examples. The U.S. environmental protection agency national conference on

retrofit opportunities for water resource protection in urban environments, Chicago, IL, July 1999, pp 13–117
Dolnicar S, Hurlimann A (2009) Drinking water from alternative water sources: differences in beliefs, social norms and factors of perceived behavioural control across eight Australian locations. Water Sci, Technol, p 1433
Durrans SR (2003) Stormwater conveyance modeling and design. Haestad Press, Waterbury, CT
Duan HD, Shibayama M (2009) Studies on Hanoi urban transition in the late 20th century based on GIS/RS. Southeast Asian Stud 46, 4
EPA (2001) National coastal condition report. EPA-620/R-01/005. US environmental protection agency, Washington, DC, p 204
EPA (2007) Innovation and research for water infrastructure for the 21st century research plan. environmental protection Agency, Office of Research and Development, Washington DC, USA
Ecosystem Valuation (2007) http://www.ecosystemvaluation.org/bigpicture.htm. Accessed Feb 2010
Elliott AH, Trowsdale SA (2007) A review of models for low impact urban stormwater drainage. Environ Model Softw 22:394–405
Fardin HF, Hollé A, Gautier-Costard E, Haury J (2014) Sanitation and water management in ancient South Asia. Evolution of sanitation and wastewater technologies through the centuries, pp 43–53
Ferguson BK (2005) Porous pavements. CRC Press, p 600
FISRWG (1998) Stream corridor restoration: principles, processes, and practices. By the Federal Interagency Stream Restoration Working Group (FISRWG)(15 Federal agencies of the US gov't). GPO Item No. 0120-A; SuDocs No. A 57.6/2:EN 3/PT.653. ISBN-0-934213-59-3
Fletcher TD, Deletic A (2007) Statistical evaluation and optimisation of stormwater quality monitoring programmes. Water Sci Technol 56(12):1–9
Frazer L (2005) Paving paradise: the peril of impervious surfaces. Environ Health Perspect 113(7):A456–A462
Grigg NS (2012) Water, wastewater, and stormwater infrastructure management. CRC Press, p 365
Guo Y, Baetz BW (2007) Sizing of rainwater storage units for green building applications. J Hydrol Eng 12(2):197–205
Hardy MJ, Kuczera G, Coombes PJ (2005) Integrated urban water cycle management: the urban cycle model. Water Sci Technol 52(9):1–9
Hatt B, Deletic A, Fletcher T (2004) Integrated stormwater treatment and re-use systems—inventory of Australian practice technical report. Research Centre for Catchment Hydrology, Melbourne
Heaney JP, Sample D, Wright L (2001) Geographical information systems decision support systems, and urban management. US Environmental Protection Agency
Hoan VQ, Edwards P (2005) Wastewater reuse through urban aquaculture in Hanoi, Vietnam: status and prospects. In: Costa-Pierce B, Desbonnet A, Edwards P, Baker D (eds) Urban aquaculture. CABI Publishing, Wallingford, pp 45–59
Holland AF, Sanger DM, Gawle CP, Lerberg SB, Santiago MS, Riekerk GHM, Zimmerman LE, Scott GI (2004) Linkages between tidal creek ecosystems and the landscape and demographic attributes of their watersheds. J Exp Mar Biol Ecol 298:151–178
Homer C, Huang C, Yang L, Wylie B, Coan M (2004) Development of a 2001 national land-cover database for the United States. Photogrammetric Engineering & Remote Sensing 70(7):829–840
Huang J, Du P, Ao C, Lei M, Zhao D, Ho M, Wang Z (2007) Characterization of surface runoff from a subtropics urban catchment. J Environ Sci 19(2):148–152
IPCC (Intergovernmental Panel on Climate Change) (2007) Climate change, 2007. The physical science basis. In: Solomon S, Qin D, Manning M, Chen Z, Marquis M, Averyt K, Tignor MMB, Miller HL, Chen Z (eds) Contribution of working group I to the fourth assessment report of the intergovernmental panel on climate change. Cambridge University Press, Cambridge, UK and New York, NY, USA, p 996
Kahinda J, Lillie E, Taigbenu A, Taute M, Boroto R (2008) Developing suitability maps for rainwater harvesting in South Africa. Phys Chem Earth 33(8–13):788–799. Weather and climate extremes in a changing climate. Regions of focus: North America, Hawaii, Caribbean, and U.S. Pacific Islands. In: Karl TR, Meehl GA, Miller CD, Hassol SJ, Waple AM, Murray WL (eds) A report

by the U.S. climate change science program and the subcommittee on global change research, Washington, D.C, 162 pp

Karl TR, Meehl GA, Miller CD, et al (eds) (2008) Weather and climate extremes in a changing climate: regions of focus: North America, Hawaii, Caribbean, and U.S. Pacific Islands. Report by the U.S. Climate Change Science Program, pp 180

Kloss C (2006) Rooftops to rivers: green strategies for controlling stormwater and combined sewer overflows. National Resources Defense Council (NRDC). https://www.nrdc.org/sites/default/files/rooftops.pdf Retrieved from July 2025

Koutsoyiannis D, Zarkadoulas N, Angelakis AN, Tchobanoglous G (2008) Urban water management in Ancient Greece: legacies and lessons. J Water Resour Plan Manag 134(1):45–54

Kumar M, Agarwal A, Bali R (2008) Delineation of potential sites for water harvesting structures using remote sensing and GIS. J Indian Soc Remote Sens 36(4):323–334

LeRoy PN, Bledsoe BP, Cuhaciyan CO (2006) Hydrologic variation with land use across the contiguous United States: geomorphic and ecological consequences for stream ecosystems. Geomorphology 79(3):264–285

Luo P, He B, Takara K, Xiong YE, Nover D, Duan W, Fukushi K (2015) Historical assessment of Chinese and Japanese flood management policies and implications for managing future floods. Environ Sci Policy 48:265–277

MacMullen E (2007) Using benefit-cost analyses to assess low-impact developments. Presentation abstract for the 2nd national low impact development conference

Malczewski J (2004) GIS-based land-use suitability analysis: a critical overview. Prog Plan 62(1):3–65

Mbilinyi B, Tumbo S, Mahoo H, Senkondo E, Hatibu N (2005) Indigenous knowledge as decision support tool in rainwater harvesting. Phys Chem Earth 30(11–16):792–798

Mbilinyi BP, Tumbo SD, Mahoo HF, Mkiramwinyi FO (2007) GIS-based decision support system for identifying potential sites for rainwater harvesting. Phys Chem Earth Parts A/B/C 32(15–18):1074–1081

Mitchell VG, Diaper C (2005) UVQ: a tool for assessing the water and contaminant balance impacts of urban development scenarios. Water Sci Technol 52(12):91–98

Mitchell VG, Duncan H, Inman M, Rahilly M, Stewart J, Vieritz A, Holt P, Grant A, Fletcher T, Coleman J, Maheepala S, Sharma A, Deletic A, Breen P (2007) Integrated urban water modelling—past, present, and future. In: Joint 13th international rainwater catchment systems conference and the 5th international water sensitive urban design conference, conference 21–23 Aug 2007, Sydney, Australia

Mitchell V, McMahon T, Mein R (2003) Components of the total water balance of an urban catchment . Environ Manag 32:735–746. https://doi.org/10.1007/s00267-003-2062-2

Nguyen DC, Dao AD, Kim T, Han M (2013) A sustainability assessment of the rainwater harvesting system for drinking water supply: a case study of Cukhe Village, Hanoi, Vietnam. Environ Eng Res 18(2):109–114

Parikh P, Taylor MA, Hoagland T, Thurston H, Shuster W (2005) Application of market mechanisms and incentives to reduce stormwater runoff: an integrated hydrologic, economic and legal approach. Environ Sci Policy 82:133–144

Perez-Pedini C, Limbrunner JF, Vogel RM (2005) Optimal location of infiltration-based best management practices for stormwater management. J Water Resour Plan Manag 131(6):441–448

Pierpont L (2008) Simulation-optimization framework to support sustainable watershed development by mimicking the pre-development flow regime. North Carolina State University, Raleigh, NC, USA Master's Thesis

Porse EC (2013) Stormwater governance and future cities. Water 5(1):29–52

Powell LM, Rohr ES, Canes ME, Cornet JL, Dzuray EJ, McDougle LM (2005) Low-impact development strategies and tools for local governments: building a business case. Report Lid50t1. LMI Government Consulting

Ragab R, Bromley J, Rosier P, Cooper JD, Gash JHC (2003) Experimental study of water fluxes in a residential area: rainfall, roof runoff and evaporation. The effect of slope and aspect. Hydrol Process 17:2409–2422

Roesner LA, Bledsoe BP (2003) Physical effects of wet weather flows on aquatic habitats: present knowledge and research needs: Alexandria, Virginia. Water Environ Res Found. Report 00-WSM-4

Rusko M, Roman C, Dana R (2010) An overview of geographic information system and its role and applicability in environmental monitoring and process modeling. Res Papers Faculty Mater Sci Technol Slovak Univ Technol 18(29):91–96

Sanger D, Blair A, DiDonato G, Washburn T, Jones S, Chapman R, Bergquist D, Riekerk G, Wirth E, Stewart J, White D, Vandiver L, White S, Whitall D (2008) Support for integrated ecosystem assessments of NOAA's National Estuarine Research Reserves System (NERRS). The impacts of coastal development on the ecology and human well-being of tidal creek ecosystems of the US Southeast, vol I. NOAA Technical Memorandum NOS NCCOS 82, p 76

Schueler TR (1994) Review of pollutant removal performance of stormwater ponds and wetlands. Watershed Protect Tech 1(1):17–18

Semadeni-Davies A, Hernebring C, Svensson G, Gustafsson L (2008) The impacts of climate change and urbanization on drainage in Helsingborg, Sweden: suburban stormwater. J Hydrol 350:114–125

Shipton MD, Somenahalli SVC (2010) Locating, appraising, and optimizing urban stormwater harvesting sites (viewed 12 Feb 10) http://www.esri.com/news/arcnews/spring10articles/locating-appraising.html

Shrestha A, Weesakul S, Babel MS (2015) Designed intensity-duration-frequency (IDF) curves under climate change condition in urban area. In: THA 2015 international conference on climate change and water & environment management in monsoon Asia, 28–30 Jan, Bangkok, Thailand

Slonecker ET, Jennings D, Garofalo D (2001) Remote sensing of impervious surface. A review. Remote Sens Rev 20(3):227–255

Sundberg C, Svensson G, Söderberg H (2004) Re-framing the assessment of sustainable stormwater systems. Clean Technol Environ Policy 6(2):120–127

Thurston HW (2006) Opportunity costs of residential best management practices for stormwater runoff control. J Water Resour Plan Manag 132(2):89–96

Tunnel Talk (2011). http://www.tunneltalk.com/Bangkok-Thailand-Dec11-Floods-bring-forward-major-mitigation-plans.php. UNDESA (Department of Economic and Social Affairs Population Division), 2014

The World Population Situation in 2014: The Concise Report. UNDESA, New York

UNDP (United Nations Development Program) (2006) Human development report 2006. Beyond scarcity: power, poverty and the global water crisis. UNDP, New York

USEPA (2002) U.S. Environmental Protection Agency (USEPA). 2000 National Water Quality Inventory. EPA-841-R-02-001. Office of Water, Washington, DC, p 207. United Nations Agenda 21, 1992

United Nations, Department of Economic and Social Affairs, Population Division (2014) World urbanization prospects: The 2014 Revision, Highlights. ST/ESA/SER. A/352

Vargas D (2009) Rainwater harvesting: a sustainable solution to stormwater management. Diss. The Pennsylvania State University

Villarreal EL (2004) Inner city stormwater control using a combination of best management practices. Ecol Eng 22(4–5):279–298

Wagner I, Izydorczyk K, Drobniewska A, Fratczak W, Zalewski M (2007) Inclusion of ecohydrology concept as integral component of systemic urban water resources management. The city of Lodz, case study, Poland. Scientific Conference SWITCH in Birmingham and New Directions in IURWM, vol 18530. SWITCHGOCE, Paris

Weng Q (2001) Modeling urban growth effects on surface runoff with the integration of remote sensing and GIS. Environ Manag 28(6):737–748

Werner AD, Alcoe DW, Ordens CM, Hutson JL, Ward JD, Simmons CT (2011) Current practice and future challenges in coastal aquifer management: flux-based and trigger-level approaches with application to an Australian case study. Water Resour Manag 25(7):1831–1853

Wilson JP, Mitasova H, Wright DJ (2000) Water resource applications of geographic information systems. Urisa J 12(2):61–79

Wong T, Brown R (2008) Transitioning to water sensitive cities: ensuring resilience through a new hydro-social contract. In: 11th international conference on urban drainage, Edinburgh, Scotland, UK

Woodward RT, Kaiser RA (2002) Market structures for U.S. water quality trading. Rev Agric Econ 24:366–383. www.epa.gov/caddis/ssr_urb_is1.html

Wynn TM (2004) The effects of vegetation on streambank erosion virginia polytechnic institute and State University, Blacksburg, Virginia. (Ph.D. Dissertation) http://scholar.lib.vt.edu/theses/available/etd-05282004-11564

Yang SL, Milliman JD, Li P, Xu K (2011) 50,000 dams later: erosion of theYangtze River and its delta. Global Planet Change 75:14–20

Chapter 7
Numerical Modeling and Simulation for Water Management

7.1 Background

Management of water resources is becoming increasingly complex (Ma et al. 2015). Demand for water supply has increased, and the need for new sources, conservation and reuse is rising. These dynamics emphasize complex interrelationships in the water management cycle, such as water source management, supply, demand, processing, reuse and resupply. The computer model has been successfully used to reproduce most of the key features of the hydrological system in order to better understand how the system functions and how it responds to different stresses (Mishra et al. 2017; Emam et al. 2016; Wurbs 1994). A model is a representation of an object, process or set of objects and processes useful for a particular purpose. Computer models help to evaluate how change in land use impacts outflow for the use in designing drainage systems. Use of simulation models to estimate variables such as floods is rapidly spreading among researchers. Water resource modeling began in the late nineteenth century as a means of addressing the design issues of urban sewerage, land reclamation drainage system and reservoir spillways (Todini 1988). Computer models were developed for decision-makers to manage and protect water resources. Decision-makers and model users play a major role in selecting, using and interpreting model results. Use of simulations models is not limited to the research field, largely used in the private, state and federal departments as well (Landström et al. 2011). Private consulting firms use a model to help clients determine the problem, such as what environmental benefits can be gained by introducing various measures, analysis of costs associated with those measures, etc. It is used as a decision-making tool to analyze the economic and environmental constraints and to support producers in determining problems such as environmental benefits and associated costs. The general public wants to know more about the risks of specific practices and management decisions.

With rapid urbanization and climate change, stormwater managers are increasingly faced with complex problems related to the design, construction, operation

B. K. Mishra et al., *Sustainable Solutions for Urban Water Security*,
Water Science and Technology Library 93,
https://doi.org/10.1007/978-3-030-53110-2_7

and maintenance of various rainwater infrastructures. Recently, ecosystem-based stormwater management systems are largely emphasized with groundwater recharge, maintenance of natural flow, downstream effects, water quality, etc. With the advent of computer systems and tools such as GIS and remote sensing, the computer model helps to simulate hydrology—hydraulic processes representing stormwater flow systems. The computer model makes it possible to test the effectiveness of different alternative stormwater management measures by simulating the water quantity and quality values at different locations. These models are based primarily on a set of differential equations representing the physical system. They describe the rate of change of various parameters with respect to time and space. Running the model means solving these equations with boundary conditions and subsequent spatial/temporal changes in the system. Model results represent the implications of global changes (e.g., climate, land use, population, etc.) and alternative control measures. Examples include depth and flow graphs at specific locations in the network. This chapter focuses on the selection, setting and application of computer models involved in the management and protection of water resources.

7.2 Model Classifications

Dramatic increase in computer capabilities and spatial databases has made unprecedented resources available to modeler. One of the most important considerations in model development is time step used in simulation. Hydrologic models fall into one of two categories: continuous and event based. Continuous time models typically are used for long-term simulations covering many days/months/years maintaining a continuous balance of the hydrologic process. Therefore, even in the absence of precipitation, the model continues to update the water balance. Event-based models are usually short duration, covering only a single cycle of variables, such as storms and responses to such events. These models are also classified as lumped and distributed model based on the model parameters as a function of space and time and deterministic and stochastic models based on the output criteria. In lumped models, the entire river basin is taken as a single unit where spatial variability is disregarded, and hence, the outputs are generated without considering the spatial processes whereas a distributed model can make predictions that are distributed in space by dividing the entire catchment into small units, usually square cells, so that the parameters, inputs and outputs can vary spatially. Deterministic model will give same output for a single set of input values, whereas in stochastic models, different values of output can be produced for a single set of inputs. One of the most popular classification methods is based on empirical (black box) approach, conceptual approach and physically based distribution approach (Devi et al. 2015). Figure 7.1 provides an illustration for different modeling approaches for estimating design flood which is largely useful in design of drainage structures.

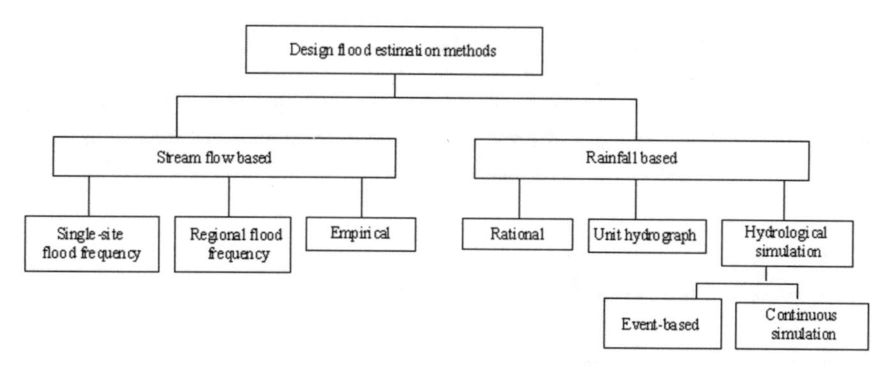

Fig. 7.1 Different modeling approaches for estimating design flood, useful for designing drainage structures

7.2.1 *Black Box Models*

There are many practical situations where the main concern is only to make an accurate river flow at a specific watershed location. In such situations, hydrologists may prefer not to spend much time and required to develop and implement conceptual or physically based models. Instead, a simpler system theoretic model, also called as black box model, is implemented. In these models, a relationship between input and output is established without taking into account the internal structure of the physical process in detail. At the point of interest, there are many conventional ways to associate the temporal and spatial characteristics of precipitation with the corresponding flow. Unit hydrograph method is an example of this method. Statistical methods use regression models, and correlation models are used to find functional relationships between inputs and outputs. The method based on the statistical approach uses classical statistical techniques to analyze historical data for the purpose of developing a method for flood prediction.

7.2.2 *Conceptual Models*

Continuous simulation is a deterministic method and involves simulating the complete time series of rainfall, runoff and streamflow. Continuous simulation includes the use of a water balance model that is generally calibrated to stream gauge data. Since the continuous model simulates the complete time series of runoff and river flow, stochastic assumptions and initial conditions related to the simulation of individual flood events are avoided. The conceptual model provides daily, monthly or seasonal estimates of continuous stream flow. These models use batch descriptions of parameters and state variables that represent average values across the settlement. The entire physical process in the hydrologic cycle is mathematically formulated in conceptual models that usually involve a large number of parameters. Many conceptual models with varying degrees of complexity have been developed. The tank model consists of a series of vertically arranged reservoir tanks, and the belt-like structure representing a zonal structure of groundwater in the objective catchment is popular conceptual model among the hydrological models (Phuong et al. 2018).

7.2.3 Physically Based Distributed Models

Physically distributed models are based on the physics of hydrological processes that control basin responses and describe these processes using physical equations. These laws are expressed in the form of partial differential equations in space and time. In discrete form, these equations are represented as difference equations, and the equations are solved at each grid point in space and time using the appropriate numeric operators. In these models, the transfer of mass, momentum and energy is dominantly solved using numerical methods such as the St. Venant equation for surface flow, the Richards equation of unsaturated region flow, and the Boussinesq equation for the groundwater flow. These models describe hydrological processes in catchments in more detail and potentially more accurately than other types. These models are extremely important in investigating the effects on climate change, land-use pattern changes, hydrological cycle by urbanization and can be used to predict outflows. The use of remote sensing data and the application of geographic information systems provides very useful input data requirements for physical hydrological models. The use of remote sensing and GIS facilitates hydrologists to respond to large, complex and spatially distributed hydrological processes. From their physical basis, such models can provide multiple outputs (e.g., river discharge, water level and evaporation loss). The physical model can overcome many flaws of the other two models due to the use of parameters with physical interpretation. It can provide large amount of information and applied in wide range of situations. The SHE/MIKE SHE model is an example of a physically distributed hydrological model (Devi et al. 2015).

7.3 Model Selection

A model is developed and evaluated for a specific purpose. These models always involve simplification of the actual system. If the simplification embedded in the model matches the purpose of the model, the usefulness of the model is enhanced, as it can focus on the most important aspect of the system. Many models are process oriented for estimating water quality and water consumption. Scientists study the physical, chemical and biological processes involved in these systems and formulate mathematical models for each component. The computer model developer combines the selected component models to form a larger model of the entire system. The model does not include all processes known to be active on large systems. Certain processes that are considered less important are omitted. Calculation and display methods are different depending on model. The model is required to generate output that can be used in the decision-making process.

Data requirement and availability needs to be considered while selecting models. High-resolution data are necessary for good prediction regardless of which model is used. The model requires different numbers and types of parameters. If possible, it is necessary to select the model according to the current purpose. The amount of computer resources needed to run the simulation is highly important in selecting a model. Detailed models that incorporate many processes in small time intervals and provide details can require great computing power. 2- and 3-dimensional models require larger computer resources. Calculation time of these hydrologic–hydraulic model can be as high as several seconds, or even days. These considerations can be very important when we incorporate soil variability and parameter uncertainty into the analysis. Mitchell et al. (2007) studied various models available to find a gap in Integrated Urban Water Management (IUWM) practice. Elliott and Trowsdale (2007) focused on the low-impact urban rainwater drainage system (LID) in order to reduce the hydrological and water quality impact of urbanization.

7.4 Calibration–Validation

Using available historical data, model calibration is carried out to adjust the model parameters to improve agreement between prediction and observation. In the absence of enough observation data, the model parameters are estimated based on the experience of modeling skill and study area. Sensitivity analysis of the model output for each parameter is crucial when assessing the effect of parameter uncertainty. If the model response is insensitive to a particular model parameter, the uncertainty in that parameter is of little concern since the model response hardly changes over the range of values of the input parameters. Distributed parameter models are generally challenging as there is no unique set of model parameters that generate a given model response. There are several parameter sets that produce the same model response or nearly the same model response. This is generally caused by a number of parameters that must be estimated and very limited information to choose from among all possible parameter sets.

The most frequently used calibration procedure is optimization of the model performance performed by comparing the observed data with the simulated data. An example of such a comparison on the calibration of the upper Bagmati River at Pandhera Dovan using AFFDEF rainfall-runoff model is reported in Fig. 7.2.

Best parameter value set can be explored using the trial and error procedure and automatic optimization techniques. An initial guess of the model parameter value is made in the beginning, a model is executed, thus obtaining a simulated data value visually/numerically compared with the corresponding observation. If the simulation is not satisfactory, the parameter value is changed and the model is executed again.

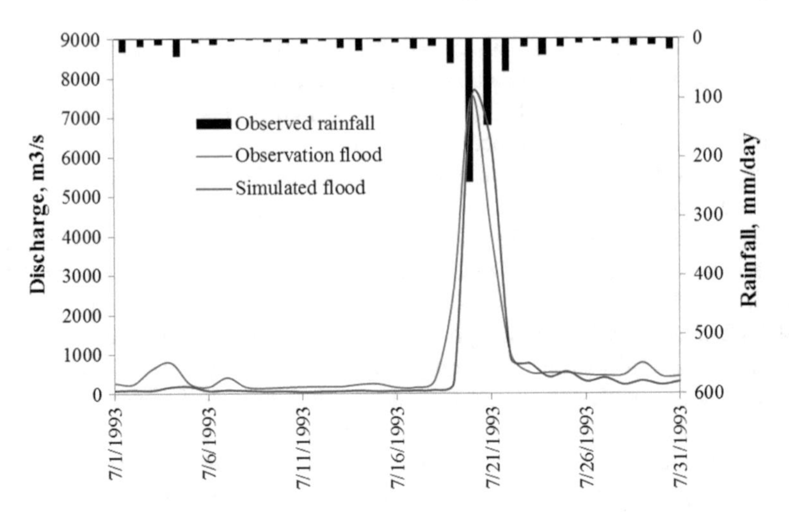

Fig. 7.2 Calibration of a rainfall-runoff model through comparison of observed and simulated data

The simulation is repeated until satisfactory solution is obtained. Model users have greater control in the trial and error procedure. However, the trial and error procedure takes a considerable amount of time with the increases in model parameters. Due to the complexity in model calibration with greater parameters, a simple model with few parameters was largely used mainly in the past.

Automatic procedures for parameter calibration are increasingly used with the advent of personal computers. Since millions of trials can be made in a relatively short time, a model containing several parameters can be quickly calibrated. In the automatic calibration, it is necessary to identify a numerical criteria to evaluate the goodness of model performance that should be executed by the algorithm. Such a numerical criteria is identified by devising an objective or loss function. For illustration, in the case of the rainfall-runoff model, the objective function compares the simulated and observed hydrographs, or the simulated characteristics of the hydrograph, such as peak and water volume.

Validation is well known in hydrology and environmental modeling and is commonly used to analyze the performance of simulations and/or predictive models. User should be careful in a sense that calibration procedure train the model with pattern of observation data in a particular context. However, when applied for future simulation, the conditions may be significantly different from those mentioned in the calibration. This is a practical problem and can adversely affect the reliability of engineering design. This problem can be mitigated by testing the calibrated model with extended sample. In several cases, it may not be possible due to limited observation data. Therefore, before using in simulation, it is recommended to test the model and check its performance in real-world applications. These test steps are called validation. An easy and intuitive way to perform a hydrological model validation is to use the so-called split sample procedure. The observation data are divided into two groups: one group is used for calibration and the other group is used to test

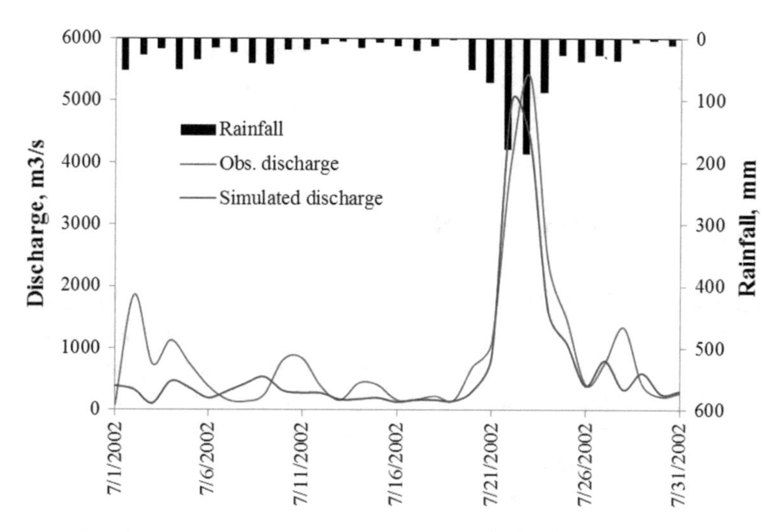

Fig. 7.3 Validation of rainfall-runoff model through comparison of observed and simulated discharge data

the model by matching the real-world situation, i.e., by running out-of-the-sample applications. Figure 7.3 shows the verification of the 2002 flood event using the calibrated AFFDEF rainfall-runoff model at Pandhera Dovan hydrological station along Bagmati River, Nepal.

7.5 Quantifying Uncertain Future

The future climate, which have major role in determining water quantity and quality, is largely stochastic phenomena. Therefore, uncertainty with the future climate should be incorporated into the modeling work while assessing the future water security. Such consideration helps in formulating realistic decision in an uncertain environment. The hydrological model can help in understanding what happens in the physical world if we adopt a specific decision strategy. Water management framework is generally formulated in a deterministic way, where reference values (for example, average, best case or worst case values) are used for key factors such as population growth, demand and availability of water in a changing context (e.g., climate change, land-use changes) although these values are largely of uncertain nature.

Uncertainty can be dealt by quantifying it in a variety of ways. In particular, the quantification of uncertainty in the water resource planning can help to improve the reliability of selected water management strategies, reduce implementation costs and allow water managers to adapt more effectively to unexpected changes in the situation. For example, the increase of the peak discharge leads to a substantial increase in the water level, and thus a considerable increase of the flood damage. The implications of possible changes in climatic conditions on the design of the flood protection structure become crucial when the change in peak discharge is calculated back for specific return period.

7.6 Open Access Data and Software

Data, knowledge and methods are developed at various level (e.g., global, national, provincial) to improve water security. It is necessary to share these data, knowledge and methods with policy-makers, planners and stakeholders to mitigate various water security problems (e.g., flood, drought, water pollution). There is need for a complete and open exchange of information that is essential to the sustainable management of water resources. These require reliable, easily accessible data or open-source data and methods. Recently, a considerable number of academic journals have established policies such as open data access, support for data storage and the provision of a dedicated venue for data disclosure.

The data such as precipitation, river flows, groundwater levels and water quality are collected by a large number of agencies and organizations using tools such as remote sensing. The data such as digital elevation model, land use/land cover, precipitation, temperature, soil moisture are an important asset for researchers and decision-makers in analyzing water resource issues. Remote sensing or satellite-based data offers advanced technologies for developing countries that lack in establishing ground monitoring network. This huge amount of information on water resources offers a vast potential for both researches and evidence-based management of resources. The collection and analysis of data also help to the researchers and governments to connect with individual water users, to raise awareness of water management challenges and to support the transparency in water governance. There are open access tools which are available for connection and wide level of knowledge sharing such as Hydrological Predictions for the Environment (HYPE) is a widely tested open-source community for researches in hydrology, hydrological modelling and source code development for the scientists, authorities and consultancies.

The open-source community will ease the sharing and enable ensemble modelling by free access to source codes. The study from Malve et al. (2012) employed the large numbers of open access data at the European scale to estimate the agricultural nonpoint load. The areas and length of river channels in the watersheds were calculated from the river and catchment database of joint research center. Some other popular open access data platforms which play important role in managing water resources are Global Earth Observation System of Systems (GEOSS), Spatial Information Platform (SIP) under SWICH-ON project, Global Data Runoff Center (GRDC), Data Distribution Centre (DDC) of the Intergovernmental Panel on Climate Change (IPCC) and the Center for Earth Resources Observation and Science (EROS) under United States Geological Survey.

In practice, data use involves careful attention to the metadata, which is the information about the data that supports the data interoperability that allows systems to work together. The European Union has developed precise guidelines to support the exchange and interoperability of the collected geospatial data through Infrastructure for Spatial Information in Europe (INSPIRE) and encourages the data accessibility though the open data portal. A more difficult challenge is posed by legal restrictions and the ownership, which differ across and within the countries. Few agencies also collect data with partially recovering their costs through the sale of data as products, which mean that the only some of the data products available freely and openly. These restrictions of data in water resources management are real barrier for researchers and limits the knowledge sharing and potentially hinders the development of tools but even within the scientific community, attitudes toward data sharing commonly varies.

7.7 Applications of Remote Sensing and GIS

Direct measurement of water resources data is always good, but in most of the cases, it is not possible at desired time and location. Reliable prediction of runoff volume and rate from the land surface to rivers is difficult and time consuming. However, this information is necessary to deal with river basin management issues. Conventional models require considerable hydrological and meteorological data. Collection of these data is expensive, time-consuming and difficult process. In order to understand hydrology of an area with sparse data availability, image processing technology provides better insights into the distribution of physical characteristics of the basin. New measurement methods (e.g., photographic system) help to evaluate areal distribution of various input variables. Remote sensing technology combined with geographical information system (GIS) can augment the conventional methods to a great extent in rainfall-runoff studies (Regan and Jackson 1980).

Remote sensing is the science that acquires information about objects or regions without making physical contact. The electromagnetic radiation, which is reflected or emitted from an object, is the usual medium to carry information in remote sensing. The role of remote sensing in runoff calculation is generally to provide the source of input data to help estimate coefficients and model parameters. Information such as land use, vegetation, drainage and soil in combination with past-measured climatic parameters (precipitation, temperature, etc.) and terrain parameter heights, contours, and gradients for the use in rainfall-runoff models.

GIS is a tool for collecting, storing, retrieving at will, transforming and displaying spatial data from the real world for a particular set of purposes. The location data are stored in vector or raster data structure, and corresponding attribute data are stored in a set of tables related geographically to the features they describe. This is also known as geographical data structure. Firstly, in spatial data structure, there is an explicit relationship between the geometric and attributes information, so that both are always available when we work with the data. Secondly, spatial data are geo-referenced to known locations on the earth's surface. Thirdly, it is designed to enable specific geographic features and phenomena to be managed. Fourthly, spatial data are organized thematically into different layers, or themes. Streams, land uses, elevation and buildings will each be stored as a separate spatial data source, rather than trying to store them all together in one. This makes it easier to manage and manipulate the data.

Water resource development with the use of RS data and GIS in the context of difficult terrain has a great importance. Recent development of satellite sensors, advanced data capture tools, easier data delivery options has expanded the accessibility and has drastically reduced the cost of many hydrological data sets. Remote Sensing and Geographic Information System (GIS) is widely applied, and integration of both is recognized as powerful and effective tool to design and formulate strategy for stormwater management. Remote sensing effectively collects the multi-temporal, multi-spectral and multilocation data and helps in understand the changes of land use, while GIS provides platform to analyzing, displaying digital data and aid as decision support system (Weng 2001). In the urban environment, the first and foremost important selection is to identify the suitable stormwater harvesting sites is of prime importance for urban water managers or management. For the same, GIS has been recommended as a decision support system (DSS) to facilitate the identification of potential stormwater harvesting sites during the decision-making process (Mbilinyi et al. 2007). GIS also functions as a screening tool for selecting preliminary sites because it has ability to spatially analyze integrated multisource data sets (Malczewski 2004). There are extensive literature on the use of GIS for evaluating site suitability in the field of rainwater collection around the world. Modeling approach with the use of GIS to handle data from computer-based tools in order to make a user-friendly decision support, which help stakeholders involved to understand and communicate of selecting technology in modern-world stormwater planning.

7.8 Case Studies

7.8.1 Simulation of Flood Inundation in Manila Using FLO-2D Model

Flooding is a serious issue in cities experiencing rapid urban growth. Assessment of spatial and temporal distribution of flood inundation is largely carried out in planning of flood risk reduction activities. Flood inundation simulation was carried out to evaluate the flood risk by coupling hydrologic–hydraulic modeling in the Marikina–Pasig River Basin, Metro Manila. The vulnerability of cities to flooding is increasing due to rapid urbanization. The rapid urbanization of Metro Manila and the slow development of infrastructure has exerted a pernicious effect on the urban environment. The annual expected damage by flooding in Metro Manila is too high due to insufficient flood control systems (e.g., pumping stations, diversion channel) and unregulated construction. In order to set up the inundation model, major input data consisted of digital elevation model (DEM), land-use, soil, diversion channel information, meteorological data (e.g., daily rainfall) and river discharge data on a daily basis. A soil map was extracted from the FAO soil database, a DEM map with 90 m resolution was extracted from Shuttle Radar Topography Mission-NASA (SRTM), and a land-use map for the whole basin with 0.5 degree resolution (Broxton.et al. 2014) was taken from MODIS-based USGS global land cover. Many of these input data were freely accessible. During heavy storms, water from the Marikina River is diverted to Laguna Lake through the Manggahan Floodway. FLO-2D model, a two-dimensional dynamic model, enabled simulation of the flood inundation of Typhoon Ondoy in August 2009. In order to understand the flood risk in the Metro Manila city area, flood vulnerable areas were identified for the NCR (Fig. 7.4). The vulnerability map was classified from low to severe based on land use and population. The results show that population density was highest in the west of NCR, which leads to an increased vulnerability of flooding in that area. The study considered that improvement of the river systems, controlling of the floods in the upstream of the Marikina River and diversion of excess waters from the Marikina and San Juan Rivers before they enter into the Pasig River can largely protect the vulnerable area of Metro Manila city.

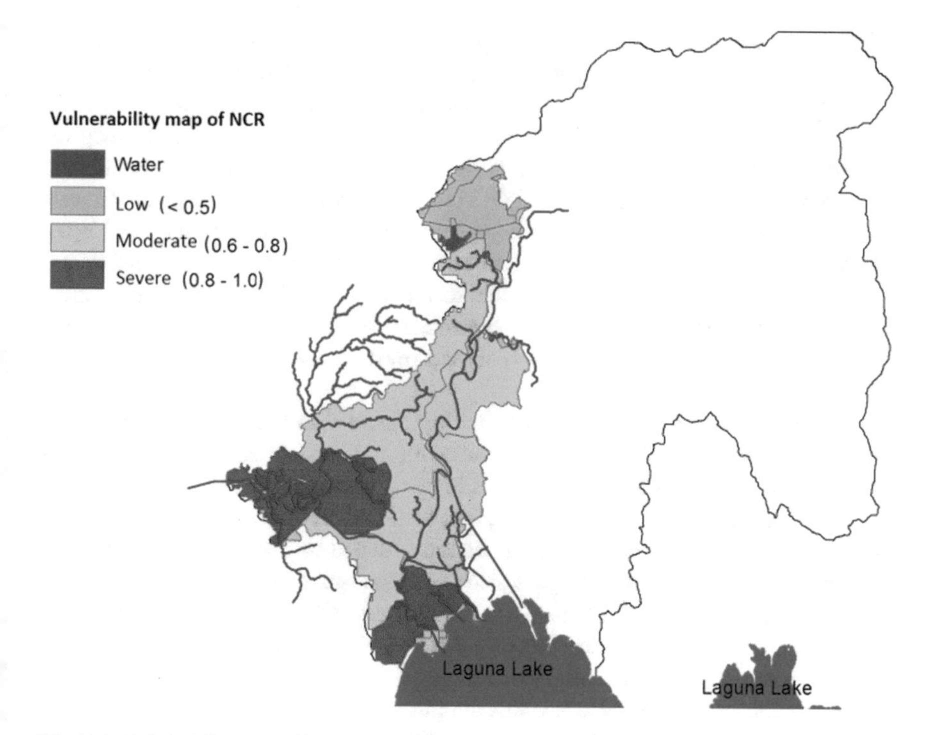

Fig. 7.4 Vulnerability map of Metro Manila city (*Source* Emam et al. 2016)

7.8.2 *Use of RS and GIS for Assessment of Land-Use Change Impacts on Runoff*

Figure 7.5 provides a flowchart on the use of remote sensing and GIS for assessing land-use change impact on runoff pattern in Bagmati River Basin, Nepal. A hydrological model, SWAT, was used to assess the runoff due to change in land use. Over the past decade, there have been substantial changes in the land use of the catchment. In the year 2001, the forest area was found to have decreased to 21.93%, while the urban area increased to 8.09% and agricultural land increased to 69% as compared to the year 1992, which had a forest area of 29.67%, an urban area of 4.28% and an agricultural area of 65.37%. As a result, the annual surface flow was found to have increased by 9.04% and the average monthly flow was found to have increased by 8.9% over the same period.

7.9 Summary

Urban water security is one of the emerging problems faced by mankind in future. Conventional hydrological data are inadequate for the purpose of design and operation of water resources systems. In such cases, remote sensing data are of great use for the estimation of the relevant hydrological variables. The synoptic concept of satellite imagery is fairly easy for identification of physical features like land use. As demonstrated, remote sensing and GIS techniques are of immense use for developing land cover maps for a hilly, inaccessible topographic country such as Nepal. Computer models together with open access data and software, remote sensing and GIS can greatly help in managing water sustainably. It will require simulation of the contributing component into computer model. Simulation model requires determining its parameters, which ultimately act as a key determination fact of the accuracy of the model result.

Urban water security is highly affected by changes in land use such as deforestation, urbanization. The increase in surface runoff and streamflow in the urban watershed is mainly due to the increase in impervious surface area and less obstruction to surface flow. The impervious surface area is directly related to urbanization whereas obstruction of surface flow is directly related to forestation. Quantitative assessment of changes in hydrologic cycle due to land-use changes in an urban basin can be very effectively and efficiently accomplished by the use of computer model, remote sensing and GIS. Remote sensing and GIS can enable to estimate the model parameters, which yield the model to simulate more or less close to real nature. Such techniques can be considered as a viable alternative or dependable support system to our conventional way of surveying, investigating, planning, monitoring, modelling, and data storing and decision-making processes.

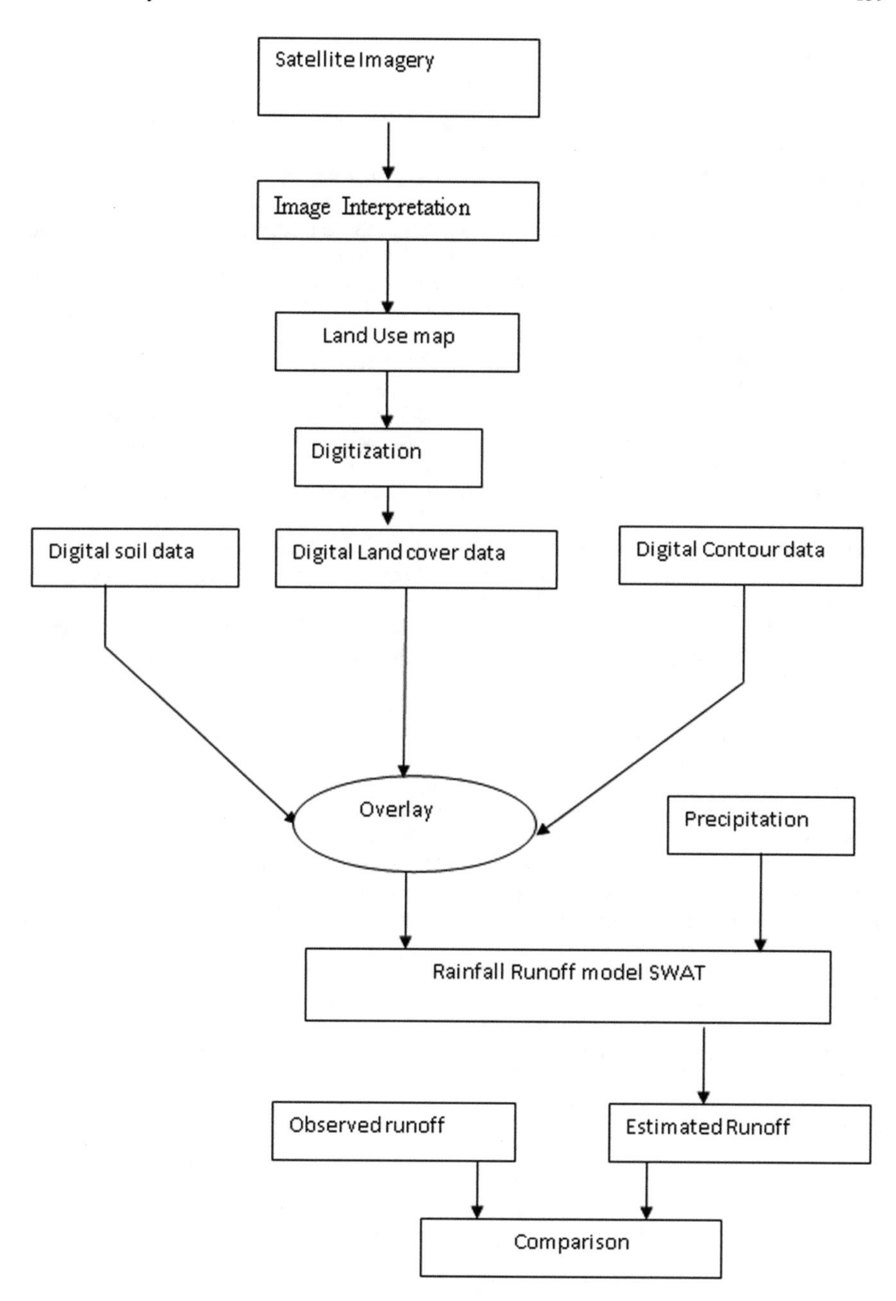

Fig. 7.5 Methodology for assessing land-use change impact on runoff using computer modeling, open access data, remote sensing and GIS

References

Broxton PD, Zeng X, Sulla-Menashe D, Troch PA (2014) a global land cover climatology using modis data. J Appl Meteor Climatol 53:1593–1605

Devi GK, Ganasri BP, Dwarakish GS (2015) A review on hydrological models. Aquat Proc 4:1001–1007

Elliott AH, Trowsdale SA (2007) A review of models of low impact urban stormwater drainage. Environ Model Softw 22:394–405. https://doi.org/10.1016/j.envsoft.2005.12.005

Emam AR, Mishra BK, Kumar P, Masago Y, Fukushi K (2016) Impact assessment of climate and land-use changes on flooding behavior in the upper Ciliwung River, Jakarta, Indonesia. Water 8, 559; https://doi.org/10.3390/w8120559

Landström C, Whatmore SJ, Lane SN (2011) Virtual engineering: computer simulation modelling for flood risk management in England. Sci Stud 24(2):3–22

Ma X, Xue X, Gonzalez-Mejia A, Garland J, Cashdollar J (2015) Sustainable water systems for the city of tomorrow—a conceptual framework. Sustainability 7:12071–12105. https://doi.org/10.3390/su70912071

Malczewski J (2004) GIS-based land-use suitability analysis: a critical overview. Prog Plan 62(1):3–65

Malve O, Tattari S, Riihimaki J, Jaakkola E, Voss A, Williams R, Baerlund I (2012) Estimation of diffuse pollution loads in europe for continental scale modelling of loads and in-stream river water quality. Hydrol Process 26:2385–2394

Mbilinyi BP, Tumbo SD, Mahoo HF, Mkiramwinyi FO (2007) GIS-based decision support system for identifying potential sites for rainwater harvesting. Phys Chem Earth, Parts A/B/C 32(15):1074–1081

Mishra BK, Regmi RK, Masago Y, Fukushi K, Kumar P, Saraswat C (2017) Assessment of Bagmati river pollution in Kathmandu valley: scenario-based modeling and analysis for sustainable urban development. Sustain Water Quality Ecol 9–10:67–77

Mitchell VG, Duncan H, Inman M, Rahilly M, Stewart J, Vieritz A, Holt P, Grant A, Fletcher TD, Coleman J, Maheepala S, Sharma A, Deletic A, Breen P (2007) State of the art review of integrated urban water models. Novatech—Lyon, France, pp 1–8

Phuong HT, Tien NX, Chikamori H, Okubo K (2018) A hydrological tank model assessing historical runoff variation in the Hieu River Basin. Asian J Water Environ Pollut 15(1):75–86

Regan RM, Jackson TJ (1980) Runoff synthesis using landsat and SCS Model. J Hydraulics Division, ASCE 106:667–678

Todini T (1988) Rainfall-runoff modeling—past, present and future. J Hydrol 100(1–3):341–352

Weng Q (2001) Modeling urban growth effects on surface runoff with the integration of remote sensing and GIS. Environ Manage 28(6):737–748

Wurbs RA (1994) Computer models for water resources planning and management. IWR Report 94-NDS-7

Chapter 8
Urban Water Governance: Concept and Pathway

8.1 Background

Water is not only essential for human well-being but also significant to the ecosystem (UN-Water 2012). Water is a vital natural resource and widely available, but the world is still facing the threat of water insecurity and scarcity (Strang 2020; Meride and Ayenew 2016). More than 1 billion of the world population is struggling to access adequate quality and quantity of drinkable water, and more than 2.6 billion lack necessary sanitation facilities (Bogardi et al. 2012). Considering the urban population growth in the last decade, access to adequate water access remains a critical issue in developing countries (Cobbinah et al. 2020; Pandey 2020). The metropolitan regions are gradually feeling the stress of rapid urbanization, the ever-growing population, and unplanned development on the local water resources. Also, water is not uniformly distributed results in physical water scarcity in one region and other struggles with floods due to the surplus of water (Cobbinah et al. 2020; Pandey 2020; Saraswat et al. 2017). Water security is becoming a concern to societies as the annual increase in global freshwater demand is continuously growing by 64 billion m^3 (cubic meter) per year, and in just last 50 years, the global water use is tripled (McDonald et al. 2014; IPCC 2014; Wada et al. 2011). The ADB AWDO (2016) report indicated that global water demand would increase by 55% by 2050 (ADB 2016; Biswas and Seetharam 2008). The reason is the current population growth rate of 1.09% globally, which is adding around 83 million people every year, increasing economic conditions, water inefficient lifestyle changes, increasing energy demands and others, exacerbating the current rate of water use.

By 2040, the parts of Asia, Africa and the Western part of South America will face the regular water stress situation (high ratio of freshwater withdrawal to supply of 40%–80%) (Cobbinah et al. 2020; World Bank 2018; Luo et al. 2015). The Asian continent is regarded as a most water-stressed continent because it has 47% of the global average of freshwater per person to serve 65% of the world's total population

B. K. Mishra et al., *Sustainable Solutions for Urban Water Security*,
Water Science and Technology Library 93,
https://doi.org/10.1007/978-3-030-53110-2_8

(Ray and Shaw 2019; Wada et al. 2013; Arnell 1999). The availability of freshwater in any region plays an important role, but at the same time, it is dependent on water use per capita. Many areas of South Asia will be under extremely high water stress, with greater than 80% of the ratio of freshwater withdrawal to supply (World Bank 2018; Pandey and Shrestha 2016). The South Asian region, with 3.5% of the world's land surface area, which is home to approximately 23.5% of world population, scored last in the Asian Water Development Outlook report 2016 (ADB 2016). This report analyzed the status of the water security situation in almost 50 countries in the Asia-Pacific region. With nearly a quarter of the world's population, the South Asian region has only 4.5% of the world's renewable freshwater resources (Ray and Shaw 2019; Pandey and Shrestha 2016). The problem of water insecurity has been generally attributed to the lack of freshwater availability, increasing pollution, insufficient funding and inadequate management of water resources, but the pressure from unplanned industrial development, rise in population growth, environmental pollution and urbanization worsening the water sustainability. This situation has been compounded by increasing variability in precipitation and global warming, due to which the region becomes extremely vulnerable to water-related natural disasters such as floods, drought and others (Pandey 2020; Ray and Shaw 2019; Saraswat et al. 2017).

Nevertheless, it is acknowledged that poor water governance in the developing countries is also another significant factor causing water insecurity and hampering the goal of sustainable development of the region (Ray and Shaw 2019). It is evident that all these factors deteriorate the water security in South Asia, and effective water governance holds the key to sustainability (Chan et al. 2016). In this respect, it is critical to divert attention to the water governance challenges and the pathways that shaping new water governance systems that are adaptive, sustainable and achieve water security (Pahl-Wostl 2019).

8.2 Water Governance: South Asia and India

Water resources in South Asia are under increasing pressure from population growth, urbanization, industrial growth along with socioeconomic water demand in the region (Ray and Shaw 2019; Arfanuzzaman and Dhaiya 2019). Also, the environmental pollution and variable changing climate patterns are affecting the availability of water resources, which is exacerbating the situation of water scarcity (du Plessis 2019; Mishra et al. 2017). In urban areas, the declining availability of water supplies and high water losses (commercial and physical) in the water distribution system are increasingly threatening water security (Chan et al. 2016). The urban water sector deals with the problem of obsolete and deteriorating water infrastructure, poor operation and maintenance, distribution system inadequacy and intermittent water supply (Sharma and Alipalo 2017; Bichai et al. 2015; Vairavamoorthy et al. 2008). Previous studies directly linked the water security issues with inefficient institutional policies, social problems and mismanagement (Mogomotsi et al. 2018; Chan et al. 2016), which led to the unsustainable use of water resources. This context highlights the need for strong and effective water governance in developing countries of South Asia.

India, a significant South Asian country, struggling from the issues of water insecurity on the societal, economic, environmental and changing climate dimensions (Singh et al. 2020; Dubey et al. 2020). From the perspective of the societal challenges, the urban region in the country has been experiencing rapid growth in population and urbanization (Roth et al. 2019), thus increasing water demand and pressure on existing water infrastructures. From an economic perspective, the challenges posed by rapid economic development and industrial growth caused an increase in water use and the deterioration of water piped infrastructure (Narain and Singh 2017; Tortajada 2016). On environmental problems, the high water losses (physical and administrative), which almost 25–40% in urban areas are due to leakage, poor operations and practices and mismanagement (Kumpel et al. 2017). Besides, the uncertain rainfall and climate change variability are posing risks to the sustainability of water provision. Overall, the challenges of the intermittent and questionable quality of water supply (McIntosh 2003), insufficient water pressure in the distribution system, unequal distribution of water access (Singh et al. 2018) and unpredictable services severed the sustainability challenges for the country. Achieving water security and sustainability is an immediate concern for India's government (Aayog 2018).

India's government experimented with the new forms of governance (adaptive, hybrid and polycentric) to ensure water sustainability and manage uncertainty (Araral and Wang 2013; UN-Water 2012). It well-argued in the literature that enhancing the effectiveness of urban water governance is vital to improve water security (Araral and Ratra 2016; Araral and Wang 2013; Bakker and Cook 2011; Cook and Bakker 2012). Effective urban water governance focuses on improving the efficiency of water-related institutions, adaptive system, strong public–private participation, protection of natural (beyond water) resources and leads to sustainable development (Rola et al. 2016) and considered as an essential driver for growth, opportunities and well-being (Rogers et al. 2003). The urban water governance is pertinent due to the decisions and policies designed are determinants of pillars of sustainability (society, economic and environmental). The main question here is which urban water governance pathway suits the Indian context. In this background of the Indian water governance system as a case study, this chapter argues that improving water security is directly proportional to the implementation of adaptive water governance. The chapter contributes to the current discourse of the water governance pathways to achieve water security in developing countries in support of the water goal (SDG 6) of the United Nation's sustainable development goals.

8.3 Urban Water Governance: Conceptualization

Water governance scholarships illustrated that the notion of urban water governance is interrelated to the pathways to achieve water security (Empinotti et al. 2019; Saraswat et al. 2016; Saraswat et al. 2017; Bakker and Morinville 2013; Briscoe and Malik 2009; Rogers 2006; Grey and Sadoff 2007). The inefficient water governance and ineffective policies are responsible for poor coordination between institutions, little or no incentives and improper allocation of resources. The concept of water governance states that formal actors' capacity determines the distribution of equitable water supply, design public policies and practical implementations (Rogers 2006). In 2002, Global Water Partnership (GWP) described the 'water governance' as a set of economic, social, administrative and political systems that regulate, manage and develop the water resources at diverse levels of society (Bakker and Morinville 2013; Gupta et al. 2013). GWP described urban water governance as four dimensions of social, economic, political and administrative systems that shape, develop and manage equitable water distribution at diverse levels of society (GWP 2002). This definition is descriptive and not clearly defines the policy implications. It also lacks a coherent analytic framework and a diagnostic value (Araral and Yu 2013). Another definition by the UNDP water governance facility provides more clarity on policy implications and clarifies the roles of actors and institutions involved. It argues that water governance addresses the equity, efficiency and effectiveness of water resources, services allocation and distribution, the need for integrated water management approaches, and balance water use between socioeconomic activities (UNDP 2013). On the other hand, Tropp argued that to steer water governance, the evolution of formal and informal networks, partnerships and coordination among actors plays a significant role (Tropp 2007). Wiek and Larson summarize water governance's key features as a systemic perspective and comprehensive perspective on water sustainability (Wiek and Larson 2012). In this chapter, urban water governance defined as the interaction of actors and networks arranged in different institutional frameworks of multi-level scales that focus on multifaceted issues binds with planning instruments.

To explore the different pathways to achieve water security, we further explore the urban water governance scholarship literature. Various authors defined their perspective of urban water governance pathways that develop the adaptive capacity of the system (Andrijevic et al. 2020; Pahl-Wostl et al. 2020; Bettini et al. 2015; Pahl-Wostl et al. 2012). Other urban water governance pathways are expressed and argued by different authors, such as Pahl-wostl, described the pathway as a system that implement norms, principles, rules and develops infrastructure to promote change in behaviors of actors (Pahl-Wostl 2009). Urban water governance pathways have been studied by scholars from their disciplinary orientations from sociology, political science, institutions, behavioral science, economics and international relations, which gives an interdisciplinary approach to the concept of urban water governance (Jiménez et al. 2020). To explore the pathways in detail, few scholars employ a comparative approach based on governance indicators (Water Law, Policy and Administration) and sub-indicators (Saleth and Dinar 2005). Others proposed strategies based on independent and in-depth case studies of good practices and suggested an incentive-based method as a pathway (Araral and Ratra 2016). These approaches complement and strengthen the water governance research agenda but provide governance pathways to achieve water security.

In this background, India's urban water governance is evaluated from two pathways: hybrid or collaborative (public–private partnership) (Tomo et al. 2020; Bakker and Morinville 2013; Rogers 2006) and adaptive system (Andrijevic et al. 2020; Pahl-Wostl et al. 2020; Bettini et al. 2015) in this study. The hybrid or collaborative system comprises neoliberal approach as an interactive arrangement of public and private actors to effectively manage and develop policies based on innovative approaches to achieve the goal of water security (Wu et al. 2016). Another pathway toward the adaptive system argues that the water system needs to predict, manage and cope with uncertainties at the local, national and global levels.

8.4 Urban Water Governance Pathway: Adaptive, Polycentric and Hybrid

The governance pathway of enhancing the water system's adaptive capabilities is significant in disaster risk reduction, building the resilience of the system and designing adaptation strategies and ecosystem management (Munaretto et al. 2014; Dietz et al. 2008). Adaptive water systems call for the shift from fixed rule-based water institutions to dynamic systems based on learning-based approaches and responsiveness (Andrijevic et al. 2020; Pahl-Wostl et al. 2020; Pahl-Wostl and Knieper 2014; Pahl-Wostl et al. 2013; Pahl-Wostl et al. 2007). The system's responsiveness in uncertain conditions and feedback and learning increases the adaptive capacity of the system (Rola et al. 2016; Bettini et al. 2015; Pahl-Wostl et al. 2012). As shown in Fig 8.1, the adaptive system's urban water governance pathway relies on the interaction among institutional arrangements under the adaptive cycle process (Jordan et al. 2015; Ostrom 2014). The adaptive system is considered resilient to changing climate due to its ability to handle future uncertainties and complexities of the water system (Huitema et al. 2009). The adaptive process system in the water sector increases the flexibility to absorb the shock and ability to manage uncertainty effectively (Rola et al. 2016). The fifth assessment report of IPCC (2014) concluded that the adaptation strategies need strong institutional support and community (Stakeholder) participation. Strong institutions support planning instruments design for the long term and necessitate the need for innovation to build effective water governance (Van De Meene et al. 2011; Hatfield-Dodds 2006).

General characteristics (dimension) of urban water governance's adaptive pathway described here, 'ability to learn' is to interpret and respond to the feedback received (Folke 2006). The system should learn from the experiences and routines and feedback (Huntjens et al. 2012). Second is the 'receptive and reactive approach' to predict and manage the climate variability impacts on the water sector (Diaz and Hurlbert 2013), robust and flexible institutions to strengthen the system and adapt (Huntjens et al. 2012). Another important characteristic is 'openness,' to assess new developments, understanding current practices and offers innovative solutions, which lead to the 'innovativeness' of the system. Adaptive systems are 'innovative' and able

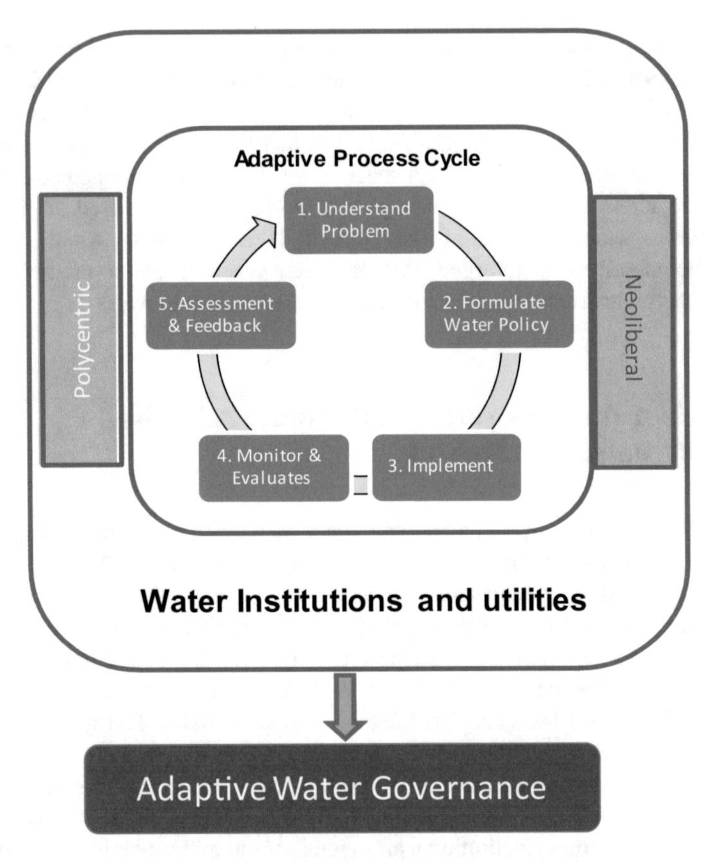

Fig. 8.1 Schematic representation of pathways (polycentric and hybrid system) to achieve the adaptive capacity (Adapted from Ayre and Nettle 2017; Pahl-Wostl et al. 2007)

to remove the silos in the system for 'integration of the fragmentation,' encourage participation and collectiveness' and 'transparency, legitimacy and equitable distribution' of water supply (Shah 2019a; Jordan et al. 2015; Huitema et al. 2009). The system has 'clearly defined boundaries' and 'informative or manage the data' for future analysis and decision-making. The polycentric and hybrid governance sought to increase the system's adaptive capacity by strengthening the institutional structure/arrangements and environment. The wide variety of water governance scholarship indicated the role of polycentricity as a pathway to increase the adaptive capacity (Carlisle and Gruby 2017; Jordan et al. 2015). The polycentric concept is based on the work of E. Ostrom (Ostrom 2014). The definition of polycentricity suggests that 'it connotes many centers of decision-making which are formally independent of each other' (Theil 2016; Jordan et al. 2018; Jordan et al. 2015; Ostrom 2014). In terms of water governance, this provides an opportunity to speculate any policy subsystem that recognized as having polycentric features, i.e., a subsystem with many independent centers of decision-making (Pahl-Wostl and Knieper 2014), yet they concur in shaping the policy. The question arises here: minimum how many formally independent decision centers are required to recognize a polycentric policy subsystem. This implies that a polycentric governance (PG) system may have fewer or larger centers of decision-making. We argue that hybrid water governance arrangements can have multiple decision-making center (e.g., the same private and public counterparts may consist of numerous actors, there may be a role for other layers of government actors, for financial actors, for local communities and other stakeholders, etc).

A recent and significant shift in urban water governance has been the engagement of different actors and networks in planning and decision-making. The hybrid or collaborative governance pathway is defined as the arrangement between formal, informal institutions partnered with private actors to achieve a common objective and advocates the limited role of the state in public management (Jensen 2017; Wu et al. 2016). All the involved actors participate and contribute to policy-making. Neoliberalism is coined with two dominating approaches; liberalization (enabling competition and the creation of markets for public goods) and privatization (a transfer of ownership and/or operations from the public to private firms) (Bakker and Morinville 2013). The private sector is expected to be innovative and efficient to survive the market competition . This infers that the efficiency orientation and innovativeness of the private sector can better meet the sustainability criteria better than the public sector (Lieberherr and Truffer 2015) is the reason for the shift. On the other hand, neoliberalism gave rise to controversies regarding the exploitation of environmental resources and the displacement of locals. The pathways, hybrid configurations, are often promoted by international organizations such as Washington Consensus and cooperation initiatives, arguing the benefit of private sector involvement in achieving water security (Gupta et al. 2013), particularly in developing countries where the sector is technology-intensive, capital intensive and complex (Biswas and Seetharam 2008). In general, the literature regards public water institutions and organizations as lacking in innovativeness (Jensen 2017). They focused on equity in water distribution, not on innovative or profit-maximizing units. In this background, the shift from primarily state-owned governance to hybrid or collaborative governance is observed in developing countries, i.e., shared ownership and operations between the public and private sectors.

8.5 Barriers to Implementation: Indian Context

The Indian urban water governance faces multifaceted challenges at the national and regional (state) level of managing water resources effectively and reducing demand (Shah 2019a). The centralized structure and inefficient institutional arrangement enforced a command–control structure that discourages transparency and reduced system effectiveness. The changing climate is inducing uncertainties in the social and ecological system, demanding an adaptive system to shape water policies (Shah 2019b; Walch 2018). In this chapter, we analyze the constraints of implementing adaptive systems in Indian water governance based on the critical aspects of governance and its pathway. One of the essential elements of adaptive governance is the characteristics of the governance regime to predict, respond and manage uncertainty caused by changing climate. Based on dimensions and water governance concept, this study characterized by the water institution's ability to learn, the responsiveness of actors and network, information access and exchange between various formal institutions, capacity building to respond to abrupt changes and equity to ensure fairness in distribution (Gupta et al. 2010; Huntjens et al. 2012). Based on the dimension mentioned, we evaluated the Indian water governance system to provide insights into the constraints to effective/adaptive water governance through pathways and how they can impact on Indian water security situation. The data are collected from the Internet, archival sources, documentary information, official contracts, and available annual reports and analyzed. Indian water institutions are classified as formal and informal institutions and distributed among national, regional, local levels. Historically, the primary responsibility of water supply and operations is with the public sector. The current majority of the Indian water governance system is under centralized, command and control structure, i.e., monocentric in nature (Shah 2019a, b). It is evaluated that the national and local level water institutions 'ability to learn' could be improved in several aspects. The reason behind the limited 'ability to learn' is the fragmentation, and the multitude of agencies, coordination and interaction is a challenge along with collaboration (Walch 2018). Recently, the Indian government attempted the problem by integrating the national water institution of a newly formed ministry 'Ministry of Jal Shakti' in 2019, which means the ministry of water power. The responsiveness of actors and networks is not effective due to the old paradigms of water management, i.e., focusing on engineering solutions for augmenting water supply. Another issue contributing unresponsiveness to actors/networks is institutional rigidity that hinders learning and lack of holistic planning.

The problem of horizontal and vertical integration is evident under the institutional structure, water management and development of water resources are entirely state subjects (Shah 2019a). There is a need for integration using a planning instrument between national and state water institutions to guide toward goal. The national government provides financial resources for the project of national interests, and state government responsibility is to manage and maintain. State water institutions are regulatory authorities, water department, public works and irrigation departments. The transparency issues of information access and exchange between different levels of institutions exist (Shah 2019a, b). Respective institutions collect data from different perspectives of water management, but sharing is not widely common, impacting long-term planning and effective decision-making. Intra- and inter collaborative efforts are required for the timely sharing of data and exchange. Another significant element of adaptive water governance is the capacity building of water-related organizations. Indian water institutions are continuously increasing their technical capacity, but innovation is not widely seen. In time to an uncertain future, it is essential to be prepared to respond, and innovation will play a significant role. To improve the innovation capacity, there is a need to focus on dynamic capabilities on the local level. Finally, equity to ensure fairness in distribution is the primary goal of the formal institution. By far in India, they are struggling with delivery water to urban sprawling (informal settlements), leakage and water losses, water theft, non-revenue water (Deshkar 2019). Overcoming barriers of addressing water losses and leakages and improving water access needs innovation and a holistic plan to achieve efficient and adequate water supply systems in the country. Also, the ability to respond is necessary to manage uncertainty posed by external factors such as climate change.

Indian water governance is a multi-level and complex system. The pathways, hybrid governance configuration (involving private actor in institutional setup and decision-making) or polycentric system configuration (multiple decision center guided by standard protocol), are recommended by governance scholarship to achieve the aim of implementing adaptive water governance (Hooda 2017). There are few examples of experimentation with a polycentric approach in India, mostly at river basin/ watershed level. Still, it is too little early to evaluate the results of its effectiveness. However, India recently started experimenting with hybrid governance configurations, the Government of India, and hails its success in Nagpur city. The public–private partnership in Nagpur exhibits the changes in institutional arrangement, organizational structure and capacity building (Deshkar 2019). Furthermore, the governance mode showed the responsiveness, improved water supply capacity and customer service by the recent improvement in water access, technologies, achieving quality of drinkable water, and continuously able to supply water to the residents but mostly in local context. There is a need for stronger regulation to steer the governance to increase the adaptive capabilities of Indian water governance. These shifts are supportive of dealing with constraints toward the adaptive governance system. Addressing the barriers herein and improving the dimension of the adaptive system mentioned in the section will enhance the strength of the urban water governance system to handle uncertainty in the future.

8.6 Summary

Generally, the water sector is considered underperforming and requires massive investment in infrastructure building and developing capacity. The complexity of the decision-making system, inability to receive timely information, conflicts of water rights (vertical and horizontal integration), poor coordination and interactions and lack of holistic planning are the reasons and important challenges of urban water governance (Cobbinah et al. 2020; Pandey 2020). This chapter explored the concept of urban water governance and its pathways to achieve adaptive water governance in India. It also contributes to the current discourse of the water governance pathways to achieve water security in developing countries in support of the water goal (SDG 6) of the United Nation's sustainable development goals. Indian water governance struggles with societal, economic, environmental and climate change challenges posing a threat to water security. The Government of India experimenting with the new forms of governance (adaptive, hybrid and polycentric) to ensure water sustainability and manage uncertainty but how to achieve that is still a considerable gap. It is argued that achieving adaptive water governance is vital to improve water security in the urban region of developing counties, including India. Adaptive urban water governance focuses on enhancing the adaptive capacity of the system, efficiency of water-related institutions, protection of natural (beyond water) resources, leads to sustainable development and considered as an important driver for growth, opportunities and well-being (Andrijevic et al. 2020; Pahl-Wostl et al. 2020; Bettini et al. 2015). In this chapter, the urban water governance is defined as the interaction of actors and networks arranged in the different institutional framework of multi-level scales focusing on multifaceted issues binds with planning instrument in the urban region. Based on the definition, a framework is designed to assess the Indian water governance system's adaptive capacity. The governance pathway of enhancing adaptive capabilities of the water system is significant in disaster risk reduction, building the resilience of the system, designing adaptation strategies and ecosystem management. Adaptive water systems call for the shift from fixed rule-based water institutions to dynamic systems based on learning-based approaches and responsiveness. Strong institutions support planning instruments design for the long term and necessitate the need for innovation to build effective water governance.

The archival reports, annual report, academic journal and secondary data analysis framework are designed. Dimensions of governance are the features that the governance regime should have to be adaptive and to predict, respond and manage uncertainty caused by external factors. The assessed dimensions are the water institution's ability to learn, the responsiveness of actors and network, information access and exchange between various formal institutions, capacity building to respond to abrupt changes and equity to ensure fairness in distribution. The polycentric and hybrid (public–private partnership) governance sought to increase the adaptive capacity of the system by strengthening the institutional structure/arrangements and environment and advanced as adaptive water governance. Several characteristics of the Indian water governance system showed a similarity with the dimension of adaptive governance. It is explored that the Indian water governance system exhibits 'ability to learn,' responsiveness, capacity, reflexivity and equity by showing the recent improvement in water access, technologies, adequate quality of drinkable water and continuously able to supply water to the residents but mostly in local context. There is a need for stronger regulation to steer the governance to increase the adaptive capabilities of the Indian water governance. However, it is analyzed that horizontal and vertical integration and coordination/communication issues between the national and local water institutions hinder the ability to handle uncertainty in Indian water regime. Addressing the constraint herein, improving on the dimension of the adaptive system mentioned in the section will enhance the strength of the urban water governance system in India to handle uncertainty in the future.

References

Aayog NITI (2018) Composite water management index. National Institution for Transforming India, GOI
ADB (2016) Asia water development outlook. Asian Development Bank, Manila
Andrijevic M, Cuaresma JC, Muttarak R, Schleussner CF (2020) Governance in socioeconomic pathways and its role for future adaptive capacity. Nat Sustain 3(1):35–41
Araral E, Ratra S (2016) Water governance in India and China: comparison of water law, policy and administration. Water Policy 18:14–31. https://doi.org/10.2166/wp.2016.102
Araral E, Wang Y (2013) Water governance 2.0: a review and second generation research agenda. Water Resour Manage 27(11):3945–3957. https://doi.org/10.1007/s11269-013-0389-x
Araral E, Yu D (2013) Comparative study of water law, policy and administration: evidence from 17 Asian countries. Water Resources Research
Arfanuzzaman M, Dahiya B (2019) Sustainable urbanization in Southeast Asia and beyond: Challenges of population growth, land use change, and environmental health. Growth and Change 50(2):725–744
Arnell NW (1999) Climate change and global water resources. Glob Environ Change. https://doi.org/10.1016/S0959-3780(99)00017-5
Ayre ML, Nettle RA (2017) Enacting resilience for adaptive water governance: a case study of irrigation modernization in an Australian catchment. Ecol Soc 22(3). https://doi.org/10.5751/ES-09256-220301
Bakker K, Cook C (2011) Water governance in Canada: innovation and fragmentation. Int J Water Resour Dev 27(2):275–289. https://doi.org/10.1080/07900627.2011.564969
Bakker K, Morinville C (2013) The governance dimensions of water security: a review. Philosophical Transactions of the Royal Society A: Mathematical, Physical and Engineering Sciences 371(2002):20130116
Bettini Y, Brown RR, de Haan FJ (2015) Exploring institutional adaptive capacity in practice: examining water governance adaptation in Australia. Ecol Soc 20(1)
Bichai F, Ryan H, Fitzgerald C, Williams K, Abdelmoteleb A, Brotchie R, Komatsu R (2015) Understanding the role of alternative water supply in an urban water security strategy: An analytical framework for decision-making. Urban Water J 12(3):175–189
Biswas AK, Seetharam KE (2008) Achieving water security for Asia. Int J Water Resour Dev. https://doi.org/10.1080/07900620701760556
Bogardi JJ, Dudgeon D, Lawford R, Flinkerbusch E, Meyn A, Pahl-Wostl C, Vörösmarty C (2012) Water security for a planet under pressure: interconnected challenges of a changing world call for sustainable solutions. Curr Opin Environ Sustain. https://doi.org/10.1016/j.cosust.2011.12.002
Briscoe J, Malik RP (2008). India's water economy: bracing for a turbulent future. Water
Carlisle K, Gruby RL (2017) Polycentric systems of governance: a theoretical model for the commons. Policy Stud J 00(00):1–26. https://doi.org/10.1111/psj.12212
Chan NW, Roy R, Chaffin BC (2016) Water governance in bangladesh: An evaluation of institutional and political context. Water, 8(9):403
Cobbinah PB, Okyere DK, Gaisie E (2020) Population growth and water supply: the future of Ghanaian cities. In Megacities and Rapid Urbanization: Breakthroughs in Research and Practice (pp. 96–117). IGI Global
Cook C, Bakker K (2012) Water security: debating an emerging paradigm. Glob Environ Change 22(1):94–102. https://doi.org/10.1016/j.gloenvcha.2011.10.011
Deshkar S (2019) Resilience perspective for planning urban water infrastructures: a case of Nagpur city. In Urban Drought (pp 131–154). Springer, Singapore
Diaz H, Hurlbert M (2013) Water governance in Chile and Canada: a comparison of adaptive characteristics. https://doi.org/10.1007/978-3-642-29831-8_11
Dietz T, Ostrom E, Stern PC (2008) The struggle to govern the commons. In: Urban ecology: an international perspective on the interaction between humans and nature. https://doi.org/10.1007/978-0-387-73412-5_40

Dubey JT, Subramanian V, Kumar N (2020) Policy interventions in achieving water security in India. In Environmental Concerns and Sustainable Development (pp. 275–291). Springer, Singapore
du Plessis A (2019) Current and future water scarcity and stress. In Water as an Inescapable Risk (pp. 13–25). Springer, Cham
Empinotti VL, Budds J, Aversa M (2019) Governance and water security: the role of the water institutional framework in the 2013–15 water crisis in São Paulo, Brazil. Geoforum 98:46–54
Folke C (2006) Resilience: the emergence of a perspective for social-ecological systems analyses. Glob Environ Change. https://doi.org/10.1016/j.gloenvcha.2006.04.002
Global Water Partnership (GWP) (2002) Introducing Effective Water Governance; GWP Technical Paper; Global Water Partnership: Stockholm, Sweden
Grey D, Sadoff CW (2007) Sink or swim? Water security for growth and development. Water policy 9(6):545–571
Gupta J, Akhmouch A, Cosgrove W, Hurwitz Z, Maestu J, Ünver O (2013) Policymakers' reflections on water governance issues 18(1)
Gupta J, Termeer C, Klostermann J, Meijerink S, van den Brink M, Jong P, Bergsma E (2010) The adaptive capacity wheel: a method to assess the inherent characteristics of institutions to enable the adaptive capacity of society. Environ Sci Policy. https://doi.org/10.1016/j.envsci.2010.05.006
Hooda SM (2017) Rajasthan water assessment: potential for private sector interventions. World Bank.
Hatfield-Dodds S (2006) The catchment care principle: a new equity principle for environmental policy, with advantages for efficiency and adaptive governance. Ecol Econ. https://doi.org/10.1016/j.ecolecon.2005.09.015
Huitema D, Mostert E, Egas W, Moellenkamp S, Pahl-Wostl C, Yalcin R (2009) Adaptive water governance: assessing the institutional prescriptions of adaptive (co-)management from a governance perspective and defining a research agenda. Ecol Soc. https://doi.org/10.5751/ES-02827-140126
Huntjens P, Lebel L, Pahl-Wostl C, Camkin J, Schulze R, Kranz N (2012) Institutional design propositions for the governance of adaptation to climate change in the water sector. Glob Environ Change. https://doi.org/10.1016/j.gloenvcha.2011.09.015
IPCC (2014) Climate change 2014 synthesis report summary chapter for policymakers. Ipcc. https://doi.org/10.1017/CBO9781107415324
Jensen O (2017) Public–private partnerships for water in Asia: a review of two decades of experience. Int J Water Resour Dev. https://doi.org/10.1080/07900627.2015.1121136
Jiménez A, Saikia P, Giné R, Avello P, Leten J, Liss Lymer B, ... Ward R (2020) Unpacking water governance: a framework for practitioners. Water 12(3):827
Jordan A, Huitema D, Schoenefeld J, van Asselt H, Forster J (2018) Governing climate change polycentrically. Governing Clim Change 3–26. https://doi.org/10.1017/9781108284646.002
Jordan AJ, Huitema D, Hildén M, Van Asselt H, Rayner TJ, Schoenefeld JJ, Boasson EL (2015) Emergence of polycentric climate governance and its future prospects. Nat Clim Change 5(11):977–982. https://doi.org/10.1038/nclimate2725
Kumpel E, Woelfle-Erskine C, Ray I, Nelson KL (2017) Measuring household consumption and waste in unmetered, intermittent piped water systems. Water Resour Res 53(1):302–315
Lieberherr E, Truffer B (2015) The impact of privatization on sustainability transitions: a comparative analysis of dynamic capabilities in three water utilities. Environ Innov Societal Trans. https://doi.org/10.1016/j.eist.2013.12.002
Luo T, Young R, Reig P (2015) Aqueduct projected water stress country rankings. World Resources Institute
McIntosh A (2003) Asian water supplies: reaching the urban poor. Asian Development Bank
McDonald RI, Weber K, Padowski J, Flörke M, Schneider C, Green PA, Montgomery M (2014) Water on an urban planet: urbanization and the reach of urban water infrastructure. Glob Environ Change. https://doi.org/10.1016/j.gloenvcha.2014.04.022
Meride Y, Ayenew B (2016) Drinking water quality assessment and its effects on residents health in Wondo genet campus, Ethiopia. Environ Syst Res 5(1):1

Mishra BKBK, Regmi RKRK, Masago Y, Fukushi K, Kumar P, Saraswat C (2017) Assessment of Bagmati river pollution in Kathmandu valley: scenario-based modeling and analysis for sustainable urban development. Sustain Water Quality Ecol 9–10:67–77. https://doi.org/10.1016/j.swaqe.2017.06.001

Mogomotsi PK, Mogomotsi GE, Matlhola DM (2018) A review of formal institutions affecting water supply and access in Botswana. Phys Chem Earth Parts A/B/C 105:283–289

Munaretto S, Siciliano G, Turvani ME (2014) Integrating adaptive governance and participatory multicriteria methods: a framework for climate adaptation governance. Ecol Soc. https://doi.org/10.5751/ES-06381-190274

Narain V, Singh AK (2017) Flowing against the current: The socio-technical mediation of water (in) security in periurban Gurgaon, India. Geoforum 81:66–75

Ostrom E (2014) A polycentric approach to climate change. Ann Econ Finance 15(1):71–108. https://doi.org/10.1596/1813-9450-5095

Pahl-Wostl C (2007) Transitions towards adaptive management of water facing climate and global change. Water Resour Manage 21(1):49–62

Pahl-Wostl C (2009) A conceptual framework for analysing adaptive capacity and multi-level learning processes in resource governance regimes. Glob Environ Change 19(3):354–365

Pahl-Wostl C (2019) The role of governance modes and meta-governance in the transformation towards sustainable water governance. Environ Sci Policy 91:6–16

Pahl-Wostl C, Knieper C (2014) The capacity of water governance to deal with the climate change adaptation challenge: using fuzzy set qualitative comparative analysis to distinguish between polycentric, fragmented and centralized regimes. Glob Environ Change 29:139–154. https://doi.org/10.1016/j.gloenvcha.2014.09.003

Pahl-Wostl C, Knieper C, Lukat E, Meergans F, Schoderer M, Schütze N, ... Thiel A (2020) Enhancing the capacity of water governance to deal with complex management challenges: A framework of analysis. Environ Sci Pol 107:23–35

Pahl-Wostl C, Lebel L, Knieper C, Nikitina E (2012) From applying panaceas to mastering complexity: toward adaptive water governance in river basins. Environ Sci Policy 23:24–34

Pahl-Wostl C, Vörösmarty C, Bhaduri A, Bogardi J, Rockström J, Alcamo J (2013) Towards a sustainable water future: shaping the next decade of global water research, 5. Curr Opin Environ Sustain § (2013). Elsevier. https://doi.org/10.1016/j.cosust.2013.10.012

Pandey CL (2020) Managing urban water security: challenges and prospects in Nepal. Environ Dev Sustain, 1–17

Pandey VP, Shrestha S (2016) Water environment in Southeast Asia: an introduction. In Groundwater Environment in Asian Cities (pp 187–191). Butterworth-Heinemann

Ray B, Shaw R (2019) Water insecurity in Asian cities. In Urban Drought (pp. 17–32). Springer, Singapore

Rogers P (2006) Water governance, water security and water sustainability. Water crisis: myth or reality, 3–36

Rogers P, Hall AW, Van de Meene SJ, Brown RR, Farrelly MA (2003) Effective water governance global water partnership technical committee (TEC). Glob Environ Change. https://doi.org/10.1016/j.gloenvcha.2011.04.003

Rola AC, Abansi CL, Arcala-Hall R, Lizada JC, Siason IML, Araral EK (2016) Drivers of water governance reforms in the Philippines. Int J Water Resour Dev 32(1):135–152. https://doi.org/10.1080/07900627.2015.1060196

Roth D, Khan MSA, Jahan I, Rahman R, Narain V, Singh AK, ... Yakami S (2019) Climates of urbanization: local experiences of water security, conflict and cooperation in peri-urban South-Asia. Climate Policy 19(sup1):S78–S93

Saleth RM, Dinar A (2005) Water institutional reforms: theory and practice. Water Policy 7(1):1–19

Saraswat C, Kumar P, Mishra BK (2016) Assessment of stormwater runoff management practices and governance under climate change and urbanization: an analysis of Bangkok, Hanoi and Tokyo. Environ Sci Policy 64. https://doi.org/10.1016/j.envsci.2016.06.018

Saraswat C, Mishra BKBK, Kumar P (2017) Integrated urban water management scenario modeling for sustainable water governance in Kathmandu Valley, Nepal. Sustain Sci 12(6):1037–1053. https://doi.org/10.1007/s11625-017-0471-z
Shah M (2019a) Crafting a paradigm shift in water. In Water Governance: Challenges and Prospects (pp. 341–368). Springer, Singapore
Shah M (2019b) Reforming india's water governance to meet 21st century challenges: practical pathways to realizing the vision of the Mihir Shah Committee. IWMI
Sharma M, Alipalo MH (2017) The Dhaka water services turnaround. Asian Development Bank
Singh S, Tanvir Hassan SM, Hassan M, Bharti N (2020) Urbanisation and water insecurity in the Hindu Kush Himalaya: insights from Bangladesh, India, Nepal and Pakistan. Water Policy 22(S1):9–32
Singh NP, Anand B, Khan MA (2018) Micro-level perception to climate change and adaptation issues: A prelude to mainstreaming climate adaptation into developmental landscape in India. Nat Hazards 92(3):1287–1304
Strang V (2020) The meaning of water. Routledge
Tomo A, Mangia G, Hinna A, Pellegrini MM (2020) Making collaborative governance effective: a case study on the pathway to successful public-private interactions. International Journal of Public Sector Performance Management 6(1):36–55
Tortajada C (2016) Policy dimensions of development and financing of water infrastructure: The cases of China and India. Environ Sci Policy 64:177–187
Tropp H (2007) Water governance: trends and needs for new capacity development. Water Policy 9(S2):19–30
UNDP (2013) User's guide on assessing water governance. Oslo: United Nations Development Programme
UN-Water (2012) WWDR4: managing water under uncertainty and risk. The United Nations World Water Development Report 4
Van de Meene SJ, Brown RR, Farrelly MA (2011) Towards understanding governance for sustainable urban water management. Glob Environ Change. https://doi.org/10.1016/j.gloenvcha.2011.04.003
Vairavamoorthy K, Gorantiwar SD, Pathirana A (2008) Managing urban water supplies in developing countries—Climate change and water scarcity scenarios. Phys Chem Earth Parts A/B/C 33(5):330–339
Wada Y, van Beek LPH, Bierkens MFP (2011) Modelling global water stress of the recent past: on the relative importance of trends in water demand and climate variability. Earth Syst Sci, Hydrol. https://doi.org/10.5194/hess-15-3785-2011
Wada Y, Van Beek LPH, Wanders N, Bierkens MFP (2013) Human water consumption intensifies hydrological drought world wide. Environ Res Let. https://doi.org/10.1088/1748-9326/8/3/034036
Walch C (2018) Adaptive governance in the developing world: disaster risk reduction in the State of Odisha, India. Clim Dev. https://doi.org/10.1080/17565529.2018.1442794
Wiek A, Larson KL (2012) Water, people, and sustainability—a systems framework for analyzing and assessing water governance regimes. Water Resour Manage 26(11):3153–3171
World Bank (2018) Water scarce cities: thriving in a finite world. Int Bank Reconstr Dev
Wu X, Schuyler House R, Peri R (2016) Public-private partnerships (PPPs) in water and sanitation in India: lessons from China. Water Policy 18(S1):153–176

Chapter 9
Toward Sustainable Solutions for Water Security

9.1 Background

This book has focused on the multifaceted concept of urban water security by analyzing a range of innovative practices, arguments and examples from the ground as well as from the literature, capturing both quantitative and qualitative (diverse) solution methods. The authors' research experiences on water resource management in several countries in Asia make this book a unique package for academicians, policymakers as well as non-experts, as it provides them with real-world examples of sustainable urban water resource management. The different solutions represented here deserve an equal weight when considering their application, with special consideration of better integration of the solutions. Multiple approaches may be combined to bring water security in an urban area.

In this chapter, we will highlight some key points that center around the dynamism of urban water security issues while synthesizing the main learning points. But before that, we will revisit the issue of water security needs in urban areas. In the era of 'Anthropocene', the human society faces an unprecedented challenge of increasing urban environments, which threatens the carrying capacity on a planetary scale. The result is that urban resource use has affected all parts of the world, from the high Himalayas (through tons of garbage) to the parts of the ocean floor (deposition of wastes). While one solution of it has been creating controls of such influences (e.g., by creating national parks or putting forward international laws to protect nature), quite evidently these are not enough. There are problems and conflicts within national parks (Bragagnolo et al. 2017), and political ecological forces can change the laws of nature protection for exploitative use (MLIT 2007). Hence is the need for an urgent consideration for sustainability needs of the urban areas that have, over the course of centuries, far gone past the ecological boundaries they are supposed to work within. It is very recent that sustainable urbanization has been looked at rigorously, involving different disciplines working together to seek workable answers [e.g., see Wilkingon et al. (2013)]. Such an integrative approach is particularly relevant for

B. K. Mishra et al., *Sustainable Solutions for Urban Water Security*,
Water Science and Technology Library 93,
https://doi.org/10.1007/978-3-030-53110-2_9

water resources, arguably the most precious of the natural resources in today's world (Hodgson 2004). It is estimated that by 2025, about half of the world's population will live in conditions of severe water stress in South Asia, Africa and Middle East (World Bank 2003). For security to be achieved, water resources should be thought of as a vital natural capital and a big contributor to people's well-being and this value must be incorporated in business, policy and development. Accordingly, this book brought together some of the workable answers on how this inclusion is possible. The book described these workable answers through different conceptual and methodological approaches (e.g., from stormwater management and ecosystem service approach by addressing diverse landscape elements to innovations that ensure affordability of the solutions). Below we note six salient points for attaining a sustainable solution to urban water security from the diverse methodological approaches.

9.2 Long-Term Sustainable Solutions

Water security problems in urban areas are multifaceted and thus are inherently complex (Chap. 1). Water security involves the security of many other resources such as food, energy, as well as problems such as climate change, and achievement of social and economic well-being (Chap. 2). Thus, water security represents a compound problem involving several different aspects. Sustainable solutions should be based on deeper understanding of water-related issues and could be specific for one issue such as water demand or water quality maintenance (Chaps. 3 and 4) or may be offered as a set of different solutions for the whole landscape (Chap. 5). These are needed due to the changing nature of water security issues (Chap. 1). Global models on climate change and its impacts on different ecosystems (e.g., coastal, mountains, dry areas), as well as socioeconomic indicators (e.g., human well-being, disaster reduction, poverty reduction and conservation of indigenous and local knowledge present in the urban areas) are essential for urban water security. Innovative resource management based on these tools can replace inefficient traditional engineering-based solutions that call for making dams, artificial reservoirs and pipelines. Growing costs of the creation and maintenance of such infrastructure, especially considering their short-term benefits (a dam can last up to a mere 100 years before its reservoir is eventually silted to an extent that inhibits its function), are a major concern. In such cases, investments in innovations and using innovations to the ground level can be more beneficial and cost-effective. The rapid urbanization combined with rapid economic development cities is experiencing degradation and depletion of water and other natural resources, which has detrimentally impacted human health, economic productivity, the quality of freshwater resources and ecosystems (Chap. 2). The conventional approaches of supply augmentation, overlooking demand management, little incentivization for water conservation and continuing timeworn water policies are leading us toward a scarce water future (Chap. 3). Using a numerical simulation hydrologic model, quantitative and qualitative assessment of changes in runoff due to urbanization and climate changes in a hydrological basin can be very effectively and efficiently accomplished (Chap. 7). In this context, the need of transition from conventional water management approaches toward urban water demand management is identified. Also, urban water governance is pertinent in decision-making and policies designed based on the determinants of sustainability. In this respect, it reviewed the concept of urban water governance and its pathways to achieve adaptive water governance (Chap. 8).

9.3 Interplay Between System Parts

Understanding the interplay between different parts of the earth surface-atmosphere system and planetary boundary conditions is crucial for sustainable resource management. Urban areas are highly dependent on technological solutions and structural modifications of the environment, which can be effective for water resource management occasionally such as the G-cans project in Tokyo (Chap. 6). But these technological solutions cannot always address all aspects of the problem unless they are combined with education and lifestyle, culture and (remaining) indigenous and local knowledge. Economic disentitlement from the resource is an even worse problem as it is socially created. Smart management practices (Chap. 4) can work in urban areas when combined with investment in water-related infrastructure that can ensure a water-secure future and bring positive changes in GDP per capita of a region or a country (Chap. 1). The innovative and new concepts such as Blue City Index (BCI) and Water Sensitive Cities Index (WSC) can be applied together with other interdisciplinary indices such as City Biodiversity Index (CBI) for better management of overall urban water environments. Water security issues can thus become a vehicle for holistic management of urban environments.

We also need special attention to trade-offs that come with security of resources. Water security can come at the expense of biodiversity, leading to an overall decrease of ecosystem benefits. The developing countries in general retain significant biodiversity but have poor infrastructural facilities and amenities. If developing countries approach water security via costly engineering methods, then riverine and freshwater-related biodiversity can deteriorate quickly. While water may be available with the help of technology, it may only be available to those who can afford it, and the urban poor is frequently disentitled in the process. The interdependencies between urban areas and the river basins within which these urban areas are located are critical for agriculture, green infrastructure in the watershed, institutional arrangements, biodiversity and economic structures related to water.

9.4 Conservation for Water Security: A Compound Criterion

The idea of conservation is not devoid of the disturbing fact of misunderstanding cultural traits, traditions and customs. In western conservation efforts, people were removed from the landscape to impose a primarily science-based command and control measure for national parks. The economic benefits of conservation for future generations are much more than a slogan because benefits of conserving natural capital become aggregated over time, but benefits of exploitative use cannot. Conservation of urban water environments can range from protecting forests in watersheds and undammed and restored rivers, to 'selective use' such as urban agriculture addressing 'green water' cycle, harvesting rainwater for drinking purpose rather than relying on large-scale structural measures, and the wise use of stormwater. These examples are illustrated throughout the book, explaining how natural capital can be protected in urban areas. Such local to regional conservation measures can maintain productive capacities of urban ecosystems. Urban communities are expected to receive most of these benefits of conservation and can also benefit from functionally connected regions (helping to reduce urban environmental footprints). In order to enhance the capacity of urban areas it is necessary to improve and raise knowledge for applying the latest techniques in conservation of water (Chap. 2).

To achieve urban water security, conservation *within* specific urban areas is a necessary step (Chap. 5). Conservation, however, as this book describes, has a compound characteristic. It does not only mean creating physical domains of conserved spaces, but in a greater sense also implies conservation through human behavior such as water demand (Chap. 3) and creating a hydrological cycle that is human-made, yet self-replenishing such as urban rainwater harvesting (this chapter). Conservation can also happen through management of water demand (Chap. 3), argued in favour of the transition from supply-oriented management toward demand-oriented management. Urban water demand management emphasizes the 'demand over supply' approach and effectively manages the demand determinants or via best practices for stormwater management (Chap. 6). Implementation of adaptive water governance through hybrid and polycentric arrangements shows promising results (Chap. 8).

9.5 Making Meanings of Large Data and Stress on Pragmatic Solutions

Urban areas produce an array of interactions with the environment. Dealing with complexity in urban water management involving source, supply, demands, treatment, re-use and back to source is a formidable challenge. This complexity can be better understood through numerical simulation models as well as data sources (e.g., remotely sensed data, social variables such as population, water usage, waste generation, water quality and quantity). Simulation models can act as tools to integrate different sets of complex data and speculate on system behavior (Chap. 3: future water demand evaluation of Kathmandu Valley) (see also Alcamo et al. 2007). The multi-criteria decision-making ability of such models has made them popular not only for research but also for private, state and federal sectors. They have high potential to help professionals in decision-making regarding costs and benefits of projects (Cominola et al. 2015).

Pragmatic solutions mean solutions that are economically viable especially considering global economic slump and initial investment bottlenecks including political bottlenecks. Applying frugal innovations, for example, can be a top priority pragmatic solution for water security issues in urban areas. Open access tools for sustainable water resource management can make implications of large data, knowledge and methods to be applied, but there are limitations how data are used continuously after a project is finished (long-term implications of a project) (Chap. 1). These can help achieve local water solutions in pragmatic manner, which can lead to better regional water security. However, these can work well only if synchronized with the range of other aspects discussed in the book.

Pragmatic solutions are needed together with novel thinking on water resource management. For example, an ecosystem service-oriented approach that takes landscape elements for sustainability may not be feasible in many of the present city environments (Chap. 5). Here, we need to conserve spaces that *surround* urban areas. For example, consideration of biodiverse landscapes in the outskirts of Manila or inclusion of remaining primary forest areas of the watershed for ecosystem-based town planning efforts at Aya town in Japan (Chap. 5) can offer useful indicators. These can be achieved through creating incentives to maintain the socio-ecological interactions in these areas or through command and control in areas that have precious biodiversity elements (e.g., endangered species, last old growth forest).

Unfortunately, river basin management is, in essence, still risk-based, which by default means structural adjustments for protection of basin communities. This type of risk-based approach requires billions of dollars, yet it is effective only for a limited period (around 100 years). Unfortunately, due to preference of immediate human security, urban water ecosystem restoration will remain a painful and slow task. This situation may be alleviated through the implementation of best management practices that combine traditional and infrastructure-based adjustments. With the help of a case study (Kathmandu Valley, Nepal), the effectiveness of urban water management strategies is evaluated, and result showed the strategies effectively reduce water demand, conserve local sources and increase water availability (Chap. 3). Considering the complex nature of water quality issues, it is very important to consider both short-term and long-term datasets of key drivers and pressures involved in the hydrological processes (Chap. 4). Remote sensing technology combined with geographical information system (GIS) can augment the conventional methods to a great extent in rainfall-runoff studies (Chap. 7). It also plays a significant role in the transitioning of the conventional water supply system to an effective water management system. There is a need of stronger regulation to steer the governance to increase the adaptive capabilities in Indian water governance. These shifts are supportive of dealing with constraints toward the adaptive governance system. This will enhance the adaptive system of current water management system and will enhance the strength of the urban water governance system to handle uncertainty in the future (Chap. 8).

9.6 Ways Forward and Gaps Remaining

Water security issues face numerous bottlenecks, some of which are discussed below:

9.6.1 Political Bottleneck and Lack of Integration

Politics in urban areas can significantly inhibit sustainable water management. Land-use policies are related to allocation of water resources and can foster economic policies that disregard the whole water environment as a form of natural capital. As evident from the chapters of this book, sustainable water resource management in urban areas depends on the efficient use of science and technology (monitoring and information, pollution treatment, health-related issues/sanitation), hunger and nutrition management, demand management and management of water-related disasters. These are required together with efficient and long-term funding. The presence or absence of such aspects can sometimes be country-specific. Considering these pressing challenges, in order to make land-use and climate change adaptation policies more effective at a local scale, importance of "Participatory Watershed Land-use Management" (PWLM) approach, a combination of participatory approaches and computer simulation modeling is very crucial (Chaps. 4 and 7). For example, the case of river (mis)management in Japan (Chap. 5) shows a lack of political integration regarding river basin management. Urban water demand management strategies play a significant role in the transitioning of the conventional water supply system to an effective water management system (Chap. 3). It is imperative, therefore, to think carefully on urban water governance pathways to achieve water security (Chap. 8) This book contributes to the current discourse of the water security in achieving the United Nation's Sustainable Development Goals (especially SDG 6). Today, policies that adapt to newer challenges in order to deliver sound and *smart* water resource management in urban areas are needed.

9.6.2 Education and Lifestyles

Urban lifestyles make us detached from the natural environments. Therefore, knowledge and training for integrated solutions are highly relevant today for answering to the multi-dimensional problems that urban ecosystems face (Chap. 1). This education can generate positive feedback as urban water security can uplift living standards, reduce poverty and advance knowledge. In addition to scientific training, optimization of knowledge on the riverine areas, urban biodiversity and their connection to human life is necessary. Education plays a pivotal part in the water governance, stimulating capacity, training, leadership and awareness potential.

9.6.3 Channelizing Research and Innovation into Policies More Efficiently

Technological innovations and frugal methods (Chap. 1) are seen as increasingly relevant by researchers around the world. However, such research also remains fragmentary, and the potential for environmentally sound decision-making is largely unfulfilled. Better synchronization between qualitative and quantitative data is an important step that needs to be grounded. Research, innovation and their inputs should be incorporated into adaptation strategies for addressing uncertainties in the water sector (e.g., climate change, political and economic situations). In a nutshell, trsndisciplinary research which is retrofiting in nature, i.e., where suggestions from different key stakeholders can be taken as input in different stage of model development to improve its researh output is the key forward to solve this complex issue of water security (Chap. 4).

9.6.4 Data Creation and Management

Data are always needed for non-structural measures and for rapid assessments. Remote sensing and numerical modeling are vital tools in this regard (Chap. 7). These approaches can also strengthen policy inputs at the ground level. Data management itself can be a solution for the growing complexity as well as the compound nature of urban water security. Moreover, creation and management of data can make us understand urban water issues as parts of a 'system' (e.g., hydrological system, social system, political system) and help to provide insight on different stressors that affect this system, together with the mediated effects of variables that work through nonlinear pathways.

9.6.5 Understanding on an Individual Basis/Leadership

Lack of leadership in water resource management can be a serious bottleneck. The majority of water management problems comes from lack of awareness in the common citizen about the health of our rivers, lakes, groundwater and artificial watercourses, and this applies to both developing and developed countries. Urban water environments in the developing nations are affected by point and nonpoint source pollution, while developed nations damage water resources through changing habitat types of urban water ecosystems and changing of resource and land-use rights. Too many stakeholders can mean that multiple parties take water management decisions and (therefore) no single authority can control them directly and efficiently. It is here that leadership can help to combat political bottleneck and lack of integration, market failure and unnecessary expense (Grigg 2011). Open models can be used more for information and analysis needed for understanding the urban water problems by the general that can work well with everyday decision-making, together with greater participation of local citizens in water resource management, creation of urban green spaces and voluntary cleaning of urban waterfronts.

With these points in mind, sustainable water resource management in urban areas will be developed in the coming decades as informed by success stories. It is hoped that this book helps in taking a step forward with wise development pathways for water resource management.

References

Alcamo J, Florke M, Marker M (2007) Future long-term changes in global water resources driven by socio-economic and climatic changes. Hydrol Sci J 52(2). http://doi.org/10.1623/hysj.52.2.247

Bragagnolo C, Correia R, Malhado ACM, Marins M, Ladle RJ (2017) Understanding non-compliance: local people's perceptions of natural resource exploitation inside two national parks in northeast Brazil. J Nat Conserv 40:64–76. https://doi.org/10.1016/j.jnc.2017.09.006

Cominola A, Giuliani M, Piga D, Castelletti A, Rizzoli AE (2015) Benefits and challenges of using smart meters for advancing residential water demand modeling and management: a review. Environ Model Softw 72:198–214. https://doi.org/10.1016/j.envsoft.2015.07.012

Grigg NS (2011) Leadership for sustainable water management: challenges and opportunities. Leadership and management in engineering. Am Soc Civil Eng 11(2). http://ascelibrary.org/doi/abs/10.1061/%28ASCE%29LM.1943-5630.0000111

Hodgson S (2004) Lands and water: the rights interface. Food and Agricultural Organizations of the United Nations, Rome. Retrieved 29 Oct 2017 from http://www.fao.org/3/a-y5692e.pdf

MLIT (Ministry of Land, Infrastructure, Transport and Tourism) (2007) Tateno dam to wa? (Tateno dam). http://www.qsr.mlit.go.jp/tateno/damujigyo/aramasi03.html. Accessed 10 Nov 2016 (In Japanese)

Wilkingon C, Sendstad M, Parnell S, Schewenius M (2013) Urbanization, biodiversity and ecosystem services: challenges and opportunities, 107–122. https://doi.org/10.1007/978-94-007-7088-1

World Bank (2003) Water resources sector strategy: strategic directions for World Bank engagement. World Bank, Washington D.C

Correction to: Sustainable Solutions for Urban Water Security

Correction to:
B. K. Mishra et al., *Sustainable Solutions for Urban Water Security*, Water Science and Technology Library 93, https://doi.org/10.1007/978-3-030-53110-2

The original version of the book was inadvertently published without funder information, which has now been added. The book has been updated with this change.

The updated version of the book can be found at
https://doi.org/10.1007/978-3-030-53110-2

B. K. Mishra et al., *Sustainable Solutions for Urban Water Security*,
Water Science and Technology Library 93,
https://doi.org/10.1007/978-3-030-53110-2_10

Printed in the United States
by Baker & Taylor Publisher Services